SpringerBriefs in Physics

SpringerBriefs in Physics are a series of slim high-quality publications encompassing the entire spectrum of physics. Manuscripts for SpringerBriefs in Physics will be evaluated by Springer and by members of the Editorial Board. Proposals and other communication should be sent to your Publishing Editors at Springer.

Featuring compact volumes of 50 to 125 pages (approximately 20,000–45,000 words), Briefs are shorter than a conventional book but longer than a journal article. Thus Briefs serve as timely, concise tools for students, researchers, and professionals.

Typical texts for publication might include:

- A snapshot review of the current state of a hot or emerging field
- A concise introduction to core concepts that students must understand in order to make independent contributions
- An extended research report giving more details and discussion than is possible in a conventional journal article
- A manual describing underlying principles and best practices for an experimental technique
- An essay exploring new ideas within physics, related philosophical issues, or broader topics such as science and society

Briefs are characterized by fast, global electronic dissemination, straightforward publishing agreements, easy-to-use manuscript preparation and formatting guidelines, and expedited production schedules. We aim for publication 8-12 weeks after acceptance.

More information about this series at http://www.springer.com/series/8902

Audrius Dubietis · Arnaud Couairon

Ultrafast Supercontinuum Generation in Transparent Solid-State Media

Springer

Audrius Dubietis
Laser Research Center
Vilnius University
Vilnius, Lithuania

Arnaud Couairon
Centre de Physique Théorique
Ecole polytechnique, CNRS, Institut
Polytechnique de Paris
Paris, France

ISSN 2191-5423   ISSN 2191-5431 (electronic)
SpringerBriefs in Physics
ISBN 978-3-030-14994-9   ISBN 978-3-030-14995-6 (eBook)
https://doi.org/10.1007/978-3-030-14995-6

Library of Congress Control Number: 2019934790

This Springer imprint is published by the registered company Springer Nature Switzerland AG
The registered company address is: Gewerbestrasse 11, 6330 Cham, Switzerland

Preface

Nonlinear propagation of intense femtosecond laser pulses in bulk transparent media produces a wealth of physical effects, whose combined action leads to a specific propagation regime, termed femtosecond filamentation, which in turn produces dramatic broadening of the pulse spectrum, termed supercontinuum generation. Due to low cost, compactness, efficiency, robustness, and virtual alignment insensitivity, femtosecond supercontinuum represents a unique and versatile source of coherent ultrabroadband radiation, whose wavelength range spans a considerable part of the optical spectrum. Altogether, these outstanding properties make femtosecond supercontinuum highly suitable for diverse applications in time-resolved spectroscopy, photonics, femtosecond technology, and lightwave engineering. During recent years, this research field has reached a high level of maturity, both in understanding of the underlying physics and in achievement of exciting practical results. In this book, we present the underlying physical picture and overview the state of the art of femtosecond supercontinuum generation in various transparent solid-state media, ranging from wide bandgap dielectrics to semiconductor materials and in various parts of the optical spectrum, from the ultraviolet to the mid-infrared. A particular emphasis is given to the most recent experimental developments: multioctave supercontinuum generation with pumping in the mid-infrared spectral range, spectral control, power and energy scaling of broadband radiation, and the development of simple, flexible, and robust pulse compression techniques, which deliver few optical cycle pulses and which could be readily implemented in a variety of modern ultrafast laser systems. The expected audience includes graduate students, professionals, and scientists working in business and academia, in the field of laser–matter interactions and ultrafast nonlinear optics.

Vilnius, Lithuania   Audrius Dubietis
Paris, France   Arnaud Couairon
January 2019

Acknowledgments

This book is the result of many years of collaborative research work. We thank members of our own research teams at Centre de Physique Théorique Ecole Polytechnique, CNRS, Institut Polytechnique de Paris (France), and Vilnius University Laser Research Center (Lithuania). We also cordially thank our national and international colleagues and numerous collaborators worldwide, and we wish to convey our gratitude to them through our acknowledgement to their respective institutions: Laboratoire d'Optique Appliquée, ENSTA ParisTech, Palaiseau, France; Femto-ST, Besançon, France; University Jean-Monnet, ST-Etienne, France; Louisiana State University, Baton Rouge, USA; University of Arizona, USA; INRS, Montréal, Canada; University of Insubria, Como & CNR, Italy; Texas A&M University at Qatar and Foundation for Research and Technology-Hellas, Heraklion, Greece; Moscow State University & Lebedev Institute, Moscow, Russia; Joint Institute for High Temperatures, RAS Moscow, Russia; ICFO Barcelona, Spain; University of Glasgow, UK; Lund University, Sweden; Leibniz University, Hannover, Germany; ETH Zurich, Switzerland; East China Normal University, Shanghai, China; and Tata Institute, Mumbai & ACRHEM, Hyderabad, India.

Preparation of this book was greatly facilitated by Tom Spicer, Cindy Zitter, Zach Evenson, and Femina Joshi from Springer Nature. We thank them for their friendly and professional work.

Contents

Chapter 1
Introduction

Supercontinuum (SC) generation is one of the most spectacular and visually perceptible effects produced by the nonlinear propagation of intense ultrashort laser pulse in a transparent medium. The discovery of supercontinuum generation in a bulk solid-state medium dates back to the early years of nonlinear optics, when Alfano and Shapiro reported on white light generation, produced by self-focusing of powerful picosecond pulses in a borosilicate glass sample [1]. The discovery was immediately followed by observations of spectral broadening in various crystals and glasses, confirming the universal nature of the phenomenon [2]. (See also [3] for a complete historical account on the early developments of SC generation in various optical media.)

   SC generation in bulk media constitutes a compact, efficient, low cost, highly robust, and virtually alignment-insensitive technique for the generation of coherent ultrabroadband radiation at various parts of the optical spectrum [4]. The physical picture of SC generation in transparent bulk media is unveiled in the framework of femtosecond filamentation, which provides a universal scenario of nonlinear propagation and spectral broadening of intense femtosecond laser pulses in bulk solids, liquids, and gases [5–9]. SC generation in bulk media appears to be a complex process that involves an intricate coupling between spatial and temporal effects: diffraction, group velocity dispersion, self-focusing, self-phase modulation, and multiphoton absorption or ionization. In the space domain, the interplay of these effects leads to the formation of a narrow light channel, termed a "light filament" that is able to propagate over extended distances much larger than the typical diffraction length and which leaves a narrow luminous plasma trail in its wake. In the time domain, the pulse undergoes dramatic transformations: pulse splitting or compression, pulse-front steepening, and generation of optical shocks. These transformations altogether produce a broadband, spatially, and temporally coherent emission with a low angular divergence (supercontinuum), which is accompanied by the generation of colored conical emission that is emitted at different angles with respect to the propagation

© The Author(s), under exclusive license to Springer Nature Switzerland AG 2019
A. Dubietis and A. Couairon, *Ultrafast Supercontinuum Generation
in Transparent Solid-State Media*, SpringerBriefs in Physics,
https://doi.org/10.1007/978-3-030-14995-6_1

axis, forming a beautiful array of concentric colored rings. Therefore, SC generation in bulk is markedly different from SC generation in optical fibers, where the propagation dynamics of the optical pulse is essentially one-dimensional and spectral broadening arises from soliton generation and fission due to the interplay between self-phase modulation and material dispersion [10].

SC generation with femtosecond laser pulses was first reported in 1983 [11], long before the phenomenon of femtosecond filamentation was discovered [12]. In that pioneering experiment, Fork and co-authors observed spectral broadening from the deep ultraviolet to the near infrared by focusing intense 80-fs pulses at 627 nm from the dye laser into an ethylene glycol jet [11]. Apart from large-scale spectral broadening, the authors underlined an improvement of pulse-to-pulse reproducibility and spatial uniformity of the beam, which resulted from the short duration of the input pulse; see also [13] and references therein for an account of SC generation in various solid-state and liquid media using femtosecond dye lasers.

A major breakthrough in femtosecond solid-state laser technology was inspired by the groundbreaking invention of chirped pulse amplification (CPA) technique by D. Strickland and G. Mourou [14], which was awarded the Nobel Prize in Physics in 2018. The CPA concept solved the long-standing problem of safe and efficient amplification of ultrashort optical pulses without the onset of optical damage of the amplifier material and other optical components, enabling a tremendous leap in the peak power and intensity of laser pulses. Shortly after that, the discovery of Kerr lens mode locking has led to the invention of femtosecond Ti:sapphire laser oscillator in 1991 [15]. Demonstration of the CPA technique-based regenerative amplification of the Ti:sapphire oscillator pulses, constituted a significant breakthrough in solid-state laser technology and marked a new era in femtosecond SC generation [16]. The amplified Ti:sapphire lasers outperformed then widely spread femtosecond dye lasers in all essential parameters of operation, setting a new standard for the entire femtosecond solid-state laser technology [17].

The commercial availability of novel femtosecond laser sources as combined with a growing practical knowledge of femtosecond SC generation in transparent condensed media [18–20], boosted the development of femtosecond optical parametric amplifiers (OPAs). In that regard, the SC radiation was recognized as an indispensable seeding source for these devices, which produced femtosecond pulses with unprecedented wavelength tunability well exceeding the tuning range afforded by conventional laser sources [21]. Broad spectral bandwidth and high temporal coherence of the SC radiation allowed compressibility of the pulses down to the Fourier transform limit, contributing to the invention of ultrabroadband noncollinear optical parametric amplifiers, the so-called NOPAs [22], which currently deliver few optical cycle pulses at various parts of the optical spectrum, ranging from the visible to the mid-infrared [23].

The advances in the optical parametric amplification techniques fostered exciting developments in the optical parametric chirped amplification (OPCPA). The general idea of the OPCPA was to replace the laser amplifier by the OPA and was originally proposed as an alternative to existing laser amplifiers in 1992 [24]. At present, OPCPA is deservedly regarded as an important offspring of the CPA technique, since as compared to the laser amplifier, the OPA offers the advantages of very high gain, broad amplification bandwidth, great wavelength flexibility, low thermal effects and superior intensity contrast of the amplified pulses, offering unique possibilities for the amplification of ultrashort laser pulses [25–27]. Interestingly, the first demonstrations of both, CPA and OPCPA, used an optical fiber to broaden the pulse spectrum before the amplification and compression stages. In that regard, SC generation in condensed bulk media offered a number of advantages due to its robustness and compactness, enabling to elaborate novel and compact architectures of tabletop SC-seeded OPCPA systems [28]. In particular, a considerable effort is currently directed to the development of the SC-seeded OPCPA systems that deliver intense few optical cycle pulses in the mid-infrared spectral region, see, e.g., [29], which is out of the grasp for existing mid-infrared solid-state lasers and laser amplifiers, see, e.g., [30].

These developments in turn facilitated experimental studies of SC generation in the region of anomalous group velocity dispersion (GVD) of dielectric solid-state media, yielding ultrabroadband, multioctave SC with unprecedented wavelength coverage, see, e.g., [31–33]. Moreover, using long-wavelength ultrashort pulses, SC generation was made possible in various highly nonlinear materials, such as narrow bandgap dielectric crystals, soft glasses, and semiconductors to produce octave-spanning SC spectra extending into far infrared. From a future perspective, SC generation represents one of the fundamental building blocks of the emerging third-generation femtosecond technology, which foresees boosting the peak and average powers of few optical cycle pulses simultaneously to the multiterawatt and hundreds of watts range, respectively, thereby paving the way for the generation of powerful sub-cycle pulses with full control over the generated light waves [34]. Finally, within the past decade, the term "supercontinuum generation" has been extended well beyond the optical range, to include high-order and nonperturbative nonlinear optical processes, such as high harmonic generation in the vacuum ultraviolet and X-ray ranges [35].

References

1. Alfano, R.R., Shapiro, L.: Emission in the region 4000 to 7000 Å via four photon coupling in glass. Phys. Rev. Lett. **24**, 584–587 (1970)
2. Alfano, R.R., Shapiro, L.: Observation of self-phase modulation and small-scale filaments in crystals and glasses. Phys. Rev. Lett. **24**, 592–594 (1970)
3. Alfano, R.R. (ed.): The Supercontinuum Laser Source. Springer (2006)
4. Dubietis, A., Tamošauskas, G., Šuminas, R., Jukna, V., Couairon, A.: Ultrafast supercontinuum generation in bulk condensed media (Review). Lith. J. Phys. **57**, 113–157 (2017)
5. Chin, S.L., Hosseini, S.A., Liu, W., Luo, Q., Thberge, F., Aközbek, N., Becker, A., Kandidov, V.P., Kosareva, O.G., Schroeder, H.: The propagation of powerful femtosecond laser pulses in optical media: physics, applications, and new challenges. Can. J. Phys. **83**, 863–905 (2005)
6. Couairon, A., Mysyrowicz, A.: Femtosecond filamentation in transparent media. Phys. Rep. **441**, 47–190 (2007)
7. Kasparian, J., Wolf, J.-P.: Physics and applications of atmospheric nonlinear optics and filamentation. Opt. Express **16**, 466–493 (2008)

8. Kandidov, V.P., Shlenov, S.A., Kosareva, O.G.: Filamentation of high-power femtosecond laser radiation. Quantum Electron. **39**, 205–228 (2009)
9. Chin, S.L.: Femtosecond Laser Filamentation. Springer (2010)
10. Dudley, J.M., Genty, G., Coen, S.: Supercontinuum generation in photonic crystal fiber. Rev. Mod. Phys. **78**, 1135–1184 (2006)
11. Fork, R.L., Shank, C.V., Hirlimann, C., Yen, R., Tomlinson, W.J.: Femtosecond white-light continuum pulses. Opt. Lett. **8**, 1–3 (1983)
12. Braun, A., Korn, G., Liu, X., Du, D., Squier, J., Mourou, G.: Self-channeling of high-peak-power femtosecond laser pulses in air. Opt. Lett. **20**, 73–75 (1995)
13. Wittmann, M., Penzkofer, A.: Spectral supebroadening of femtosecond laser pulses. Opt. Commun. **126**, 308–317 (1996)
14. Strickland, D., Mourou, G.: Compression of amplified chirped optical pulses. Opt. Commun. **56**, 219–221 (1985)
15. Spence, D.E., Kean, P.N., Sibbett, W.: 60-fsec pulse generation from a self-mode-locked Ti:sapphire laser. Opt. Lett. **16**, 42–44 (1991)
16. Norris, T.B.: Femtosecond pulse amplification at 250 kHz with a Ti:sapphire regenerative amplifier and application to continuum generation. Opt. Lett. **17**, 1009–1011 (1992)
17. Backus, S., Durfee, C.G., Murnane, M.M., Kapteyn, H.C.: High power ultrafast lasers. Rev. Sci. Instr. **69**, 1207–1223 (1998)
18. Brodeur, A., Chin, S.L.: Band-gap dependence of the ultrafast white-light continuum. Phys. Rev. Lett. **80**, 4406–4409 (1998)
19. Brodeur, A., Chin, S.L.: Ultrafast white-light continuum generation and self-focusing in transparent condensed media. J. Opt. Soc. Am. B **16**, 637–650 (1999)
20. Bradler, M., Baum, P., Riedle, E.: Femtosecond continuum generation in bulk laser host materials with sub-μJ pump pulses. Appl. Phys. B **97**, 561–574 (2009)
21. Cerullo, G., De Silvestri, S.: Ultrafast optical parametric amplifiers. Rev. Sci. Instr. **74**, 1–18 (2003)
22. Wilhelm, T., Piel, J., Riedle, E.: Sub-20-fs pulses tunable across the visible from a blue-pumped single pass noncollinear parametric converter. Opt. Lett. **22**, 1494–1496 (1997)
23. Brida, D., Manzoni, C., Cirmi, G., Marangoni, M., Bonora, S., Villoresi, P., De Silvestri, S., Cerullo, G.: Few-optical-cycle pulses tunable from the visible to the mid-infrared by optical parametric amplifiers. J. Opt. **12**, 013001 (2010)
24. Dubietis, A., Jonušauskas, G., Piskarskas, A.: Powerful femtosecond pulse generation by chirped and stretched pulse parametric amplification in BBO crystal. Opt. Commun. **88**, 437–440 (1992)
25. Dubietis, A., Butkus, R., Piskarskas, A.P.: Trends in chirped pulse optical parametric amplification. IEEE J. Sel. Topics Quantum Electron. **12**, 163–172 (2006)
26. Witte, S., Eikema, K.S.E.: Ultrafast optical parametric chirped-pulse amplification. IEEE J. Sel. Topics Quantum Electron. **18**, 296–307 (2012)
27. Vaupel, A., Bodnar, N., Webb, B., Shah, L., Richardson, M.: Concepts, performance review, and prospects of table-top, few-cycle optical parametric chirped-pulse amplification. Opt. Eng. **53**, 051507 (2014)
28. Budriūnas, R., Stanislauskas, T., Adamonis, J., Aleknavičius, A., Veitas, G., Gadonas, D., Balickas, S., Michailovas, A., Varanavičius, A.: 53 W average power CEP-stabilized OPCPA system delivering 5.5 TW few cycle pulses at 1 kHz repetition rate. Opt. Express **25**, 5797–5806 (2017)
29. Rigaud, P., van de Walle, A., Hanna, M., Forget, N., Guichard, F., Zaouter, Y., Guesmi, K., Druon, F., Georges, P.: Supercontinuum-seeded few-cycle midinfrared OPCPA system. Opt. Express **24**, 26494–26502 (2016)
30. Pires, H., Baudisch, M., Sanchez, D., Hemmer, M., Biegert, J.: Ultrashort pulse generation in the mid-IR. Prog. Quantum Electron. **43**, 1–30 (2015)
31. Silva, F., Austin, D.R., Thai, A., Baudisch, M., Hemmer, M., Faccio, D., Couairon, A., Biegert, J.: Multi-octave supercontinuum generation from mid-infrared filamentation in a bulk crystal. Nature Commun. **3**, 807 (2012)

32. Couairon, A., Jukna, V., Darginavičius, J., Majus, D., Garejev, N., Gražulevičiūtė, I., Valiulis, G., Tamošauskas, G., Dubietis, A., Silva, F., Austin, D.R., Hemmer, M., Baudisch, M., Thai, A., Biegert, J., Faccio, D., Jarnac, A., Houard, A., Liu, Y., Mysyrowicz, A., Grabielle, S., Forget, N., Durécu, A., Durand, M., Lim, K., McKee, E., Baudelet, M., Richardson, M.: Filamentation and pulse self-compression in the anomalous dispersion region of glasses, in Laser Filamentation, eds, pp. 147–165. CRM Series in Mathematical Physics (Springer, A. D. Bandrauk et al (2016)
33. Chekalin, S.V., Dormidonov, A.E., Kompanets, V.O., Zaloznaya, E.D., Kandidov, V.P.: Light bullet supercontinuum. J. Opt. Soc. Am. B **36**, A43–A53 (2019)
34. Fattahi, H., Barros, H.G., Gorjan, M., Nubbemeyer, T., Alsaif, B., Teisset, C.Y., Schultze, M., Prinz, S., Haefner, M., Ueffing, M., Alismail, A., Vámos, L., Schwarz, A., Pronin, O., Brons, J., Geng, X.T., Arisholm, G., Ciappina, M., Yakovlev, V.S., Kim, D.-E., Azzeer, A.M., Karpowicz, N., Sutter, D., Major, Z., Metzger, T., Krausz, F.: Third-generation femtosecond technology. Optica **1**, 45–63 (2014)
35. Zheltikov, A.: Multioctave supercontinua and subcycle lightwave electronics. J. Opt. Soc. Am. B **36**, A168–A182 (2019)

Part I
Physical Picture of Supercontinuum Generation

Chapter 2
Governing Physical Effects

During the last two decades, significant progress has been accomplished in the development of optical fibers for the generation of ultrabroadband high-brightness spectra through supercontinuum (SC) generation, see e.g., [1]. The nonlinear propagation dynamics of the optical pulse in a fiber is essentially one-dimensional since single-mode propagation over broad wavelength ranges is desired to ensure good guidance properties and high nonlinearity over extended lengths. Supercontinuum generation results from the interplay between self-phase modulation and material dispersion, which leads to spectral broadening via soliton generation and fission. Supercontinuum generation in bulk media, first reported by Alfano and Shapiro [2, 3], appears to be a more complex process that involves an intricate coupling between spatial and temporal effects. The physics of SC generation in transparent bulk media could be understood in the framework of investigations on femtosecond filamentation, which denotes a universal phenomenon of nonlinear propagation and spectral broadening of intense femtosecond laser pulses over long distances in solids, liquids, and gases [4–6]. This phenomenon was first observed in air in 1995 [7], after the chirped pulse amplification technology became available [8].

Femtosecond filamentation arises as a nonlinear propagation regime that is featured by competing physical effects, among which self-focusing, self-phase modulation, and multiphoton absorption/ionization-induced free electron plasma. To encompass the numerous observations of filaments in various propagation media, Couairon and Mysyrowicz proposed the following definition: A filament denotes "a dynamic structure with an intense core, that is able to propagate over extended distances much larger than the typical diffraction length while keeping a narrow beam size without the help of any external guiding mechanism" [5]. High intensities exceeding the ionization threshold of the propagation medium is difficult to measure directly since any optical element could be damaged in the process. However, in practise, the most striking and obvious manifestation of filament formation is the generation of an extremely broadband, spatially and temporally coherent emission with a low

© The Author(s), under exclusive license to Springer Nature Switzerland AG 2019
A. Dubietis and A. Couairon, *Ultrafast Supercontinuum Generation in Transparent Solid-State Media*, SpringerBriefs in Physics,
https://doi.org/10.1007/978-3-030-14995-6_2

angular divergence (or SC generation), accompanied by the generation of colored conical emission, i.e., broadband radiation emitted at different angles with respect to the propagation axis, forming a beautiful pattern of concentric colored rings on a screen intersecting the beam. In the case of filaments in air, the scale of experiments leading to the observation of supercontinuum generation and conical emission is typically longer than a meter and requires the input pulse energies of the order of several milijoules. In condensed media (solids and liquids), with modern ultrashort pulse laser sources, nonlinear effects responsible for filamentation and supercontinuum generation show up for the input pulse energies of several microjoules and less, and are reduced to a compact, few millimeter to few centimeter length scales owing to the higher Kerr nonlinearity of the propagation medium.

2.1 Self-focusing of Laser Beams

When an intense laser pulse propagates in a material, the response of matter to the electric field not only consists in the linear susceptibility $\chi^{(1)}$ but also in nonlinear terms in the dielectric polarization density, which can be written as a Taylor series expansion

$$\mathbf{P}(t) = \epsilon_0 \left(\chi^{(1)}\mathbf{E}(t) + \chi^{(2)}\mathbf{E}^2(t) + \chi^{(3)}\mathbf{E}^3(t) + \cdots \right), \tag{2.1}$$

where ϵ_0 denotes the permittivity in vacuum. A quasi-instantaneous response of bound electrons has been assumed, resulting in the independence of the nth order susceptibility $\chi^{(n)}$ upon frequency. In general, $\chi^{(n)}$ is a tensor of rank $n + 1$, depending on the vectorial nature of fields $\mathbf{E}$ and $\mathbf{P}$, and on the symmetries of the medium. In Eq. (2.1), we consider linearly polarized electric fields, allowing us to write the equivalent scalar relation. If we consider materials with inversion symmetry, at lowest order, the dominant nonlinear susceptibility is the cubic term. The dielectric polarization density

$$P(t) = \epsilon_0 \left(\chi^{(1)}E(t) + \chi^{(3)}E^3(t) \right) \tag{2.2}$$

is associated with a change of refractive index induced by the high intensity $I \equiv \epsilon_0 c n_0 |E|^2/2$:

$$n = n_0 + n_2 I, \tag{2.3}$$

where $n_0 = \chi^{(1)1/2}$ is the linear refractive index and n_2 is the nonlinear refractive index coefficient, linked to the third-order (cubic) optical susceptibility of the material by the relation

$$n_2 = \frac{3\chi^{(3)}}{4\epsilon_0 c n_0^2}. \tag{2.4}$$

It is positive in the transparency range of most dielectric media.

During its propagation, a laser beam induces an increase of refractive index proportional to the local intensity and thus the beam is exposed to a higher index at the center and a lower index at the edges. The effect on the beam is similar to that of a lens, enforcing the beam to self-focus. However, in contrast with linear focusing by a lens, self-focusing is a cumulative effect along propagation as it makes the beam more intense in its center, in turn enhancing the refractive index and the trend to self-focus. In the absence of any saturation effect preventing the intensity in the center of the beam to grow indefinitely, self-focusing would end up in a singularity (catastrophic collapse) at a finite propagation distance. This situation does not appear in practise since several saturation effects are able to counteract self-focusing when intensity exceeds a certain threshold. The physics of self-focusing and Kerr-induced beam collapse is however one of the first nonlinear effects that were investigated in detail, in the framework of models accounting only for diffraction and the optical Kerr effect, without saturation. An understanding of these phenomena was indeed crucial for the development of high-power lasers, so as to avoid or at least control beam self-focusing in order not to reach high intensities that potentially exceed the damage threshold of optical materials.

Self-focusing always compete with diffraction. The intensity-dependent refractive index indeed modifies the phase of the propagating beam but it is the action of diffraction which conveys the changes in the phase induced by nonlinearity and ultimately leads to self-focusing, i.e., to the decrease of the beam width along propagation. The number of transverse dimensions in which the beam can spread under the effect of diffraction is important for the occurrence of collapse.

In the particular case of a single transverse dimension, for example, in a slab waveguide geometry where one transverse dimension is guided while diffraction acts freely in the other transverse dimension, collapse never occurs. If diffraction overcomes self-focusing, the beam spreads. If self-focusing overcomes diffraction, the beam shrinks, but this process stops when the beam width is small enough and diffraction becomes again dominant, resulting in an oscillatory propagation featured by defocusing–refocusing cycles. There is an intermediate situation, where the strength of diffraction is equal to the strength of self-focusing, leading to an exactly balanced propagation in the form of a spatial soliton.

In two transverse dimensions, for a cylindrically symmetric Gaussian beam, the competition between diffraction and self-focusing is determined by the beam power. As in the one-dimensional case, it is possible to find a specific beam shape corresponding to a spatial soliton, for which diffraction and the self-focusing nonlinearity are exactly balanced. This soliton is known as the *Townes mode* [9]. It contains a well-defined power

$$P_{cr} = \frac{3.72\lambda^2}{8\pi n_0 n_2},$$
(2.5)

where λ is the laser wavelength. The Townes mode is a mathematically exact propagation invariant of a unidirectional propagation equation taking into account only the effects of diffraction and Kerr nonlinearity. Originally, it was also called a *self-trapped mode* [9].

In the frame of this ideal model, it turns out that the Townes mode is not stable. This means that if we assume that the Townes mode could be generated in the laboratory, any small perturbation of the beam amplitude or phase would likely be amplified and prevent the beam to propagate as an invariant Townes mode. This general statement may apply to several types of instability which can be investigated rigorously, see e.g., [10]. Here, an intuitive picture is sufficient to illustrate the structural instability of the Townes mode. If the beam power is above P_{cr}, the Kerr nonlinearity overcomes diffraction and the beam shrinks upon itself and undergoes a collapse singularity at a finite propagation distance. On the other hand, if the beam power is lower than P_{cr}, diffraction overcomes self-focusing and the beam will ultimately spread. Equation (2.5) therefore not only corresponds to the power of the Townes mode but also to the minimum power that a beam must exceed in order to undergo a collapse singularity. It is often (and inappropriately) called the critical power threshold for self-focusing in the literature; that is the power, when the effect of self-focusing precisely balances the diffractive spreading of the beam, even though beam self-focusing can occur below P_{cr} and is eventually dominated by beam spreading.

The shape of a Gaussian beam is close to that of the Townes beam, whose power realizes the balance between self-focusing and diffractive spreading. However, Marburger showed that any beam different from the Townes beam requires a larger power than P_{cr} to undergo a catastrophic collapse [11]. Dawes and Marburger determined by numerical simulations the critical power for a Gaussian beam [12]:

$$P_{cr}^* = \frac{3.77\lambda^2}{8\pi n_0 n_2}. \tag{2.6}$$

Their simulations showed that if the power P of a collimated input Gaussian beam exceeds P_{cr}^*, the beam will self-focus at a distance z_{sf}, called the nonlinear focus [11]:

$$z_{sf} = \frac{0.367 z_R}{\sqrt{[(P/P_{cr}^*)^{1/2} - 0.852]^2 - 0.0219}}. \tag{2.7}$$

Here, $z_R = \pi n_0 w_0^2/\lambda$ denotes the Rayleigh (diffraction) length of the input Gaussian beam of e^{-2}-radius w_0. Although Eq. (2.7) is derived in the case of continuous wave laser beams, it gives a fairly accurate approximation of the nonlinear focus of femtosecond laser pulses as well. A pulse can indeed be viewed as stacked into independent time slices, each of which corresponding to a beam carrying its own power $P(t)$. All slices whose power exceeds P_{cr}^* will self-focus at a distance given by Eq. (2.7). Thus, the shorter distance leading to an intensity peak is obtained for the central slice corresponding to the peak power of the pulse.

Figure 2.1 shows an example of the evolution of the beam radius during self-focusing of a loosely focused femtosecond-pulsed Gaussian beam with power of $\sim 5\ P_{cr}$ in water. The position of the nonlinear focus is indicated by the minimum beam radius.

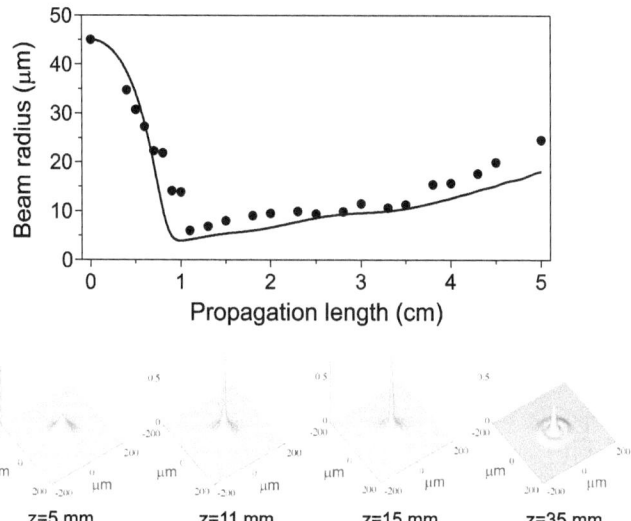

Fig. 2.1 FWHM beam radius as a function of propagation distance during self-focusing in water of a loosely focused Gaussian input beam with an FWHM diameter of 90 μm. The input beam power is $\sim$5 P_{cr}. The solid curve shows the numerical simulation, bold dots show the experimental data. The images below show fluence profiles of the beam before ($z = 5$ mm), at ($z = 11$ mm) and after ($z = 15, 35$ mm) the nonlinear focus. Adapted from [13]. Reprinted by permission from Springer Nature

2.2 Self-phase Modulation of Laser Pulses

The intensity-dependent refractive index change expressed in Eq. (2.3) not only varies in the transverse direction with the local intensity but also in time when a laser pulse with time-dependent intensity profile is considered. In the same way as the refractive index change was inducing a change of the spatial phase of the beam, in the time domain, the time dependence of the refractive index imparts a nonlinear change in the phase of the pulse

$$\phi_{nl}(t) = \frac{\omega_0}{c} n_2 \int_0^L I(t, z) \, dz, \tag{2.8}$$

where ω_0 is the carrier frequency, z is the propagation distance, and L is the length of the nonlinear medium. Equation (2.8) shows that a nonlinear phase shift is accumulated during the pulse propagation. For this reason, the effect is called self-phase modulation. The intensity $I(t, z)$ generally varies along the propagation axis z but for simplicity, we consider a nonlinear medium of length L that is short enough for the intensity variation over z to be negligible, allowing us to easily rewrite the nonlinear phase as $\phi_{nl}(t) = (\omega_0 L/c) n_2 I(t)$. This produces a frequency change $\delta\omega(t) = -\frac{d}{dt}\phi_{nl}(t)$ that results in a time-varying instantaneous frequency:

$$\omega(t) = \omega_0 + \delta\omega(t), \tag{2.9}$$

where $\delta\omega(t) = -(\omega_0 L/c)n_2 \partial I/\partial t$, giving rise to spectral broadening of the pulse. For a Gaussian laser pulse of duration t_p, the variation of the instantaneous frequency is expressed as

$$\delta\omega(t) = 4\frac{\omega_0 L}{ct_p^2}n_2 I_0 t \exp\left(-2\frac{t^2}{t_p^2}\right). \tag{2.10}$$

The effect therefore leads to the generation of new frequencies, i.e., gives rise to spectral broadening by inducing a negative shift of the instantaneous frequency at the leading (ascending) front of the pulse and a positive shift of the instantaneous frequency at the trailing (descending) front of the pulse, as schematically illustrated in Fig. 2.2. In other words, the pulse acquires a frequency modulation corresponding to the production of red-shifted spectral components at the pulse front and blue-shifted spectral components at the pulse tail.

Figure 2.3 illustrates the effect of self-phase modulation on the spectrum of a laser pulse assumed to undergo only this effect, i.e., the pulse shape is preserved along propagation. The first row displays phase modulations generated for a symmetric input Gaussian pulse. It shows that the input spectrum broadens and becomes more and more modulated as the propagation distance increases. The second row shows the spectral broadening for an asymmetric input pulse with a steeper trailing front compared to the leading front. In that case, spectral broadening is asymmetric, extending farther on the high-frequency side compared to the low-frequency side, in agreement with Eq. (2.10).

Information about the frequency content of different parts of the pulse can be obtained from the spectrograms shown in Fig. 2.4. The rows correspond to the cases of a symmetric (first row) and asymmetric pulse, as in Fig. 2.3. Compared to the symmetric distribution of frequencies for the input pulses, Fig. 2.4a and d, a frequency

Fig. 2.2 Self-phase modulation of a Gaussian pulse shown in (**a**), which produces a variation of the instantaneous frequency shown in (**b**)

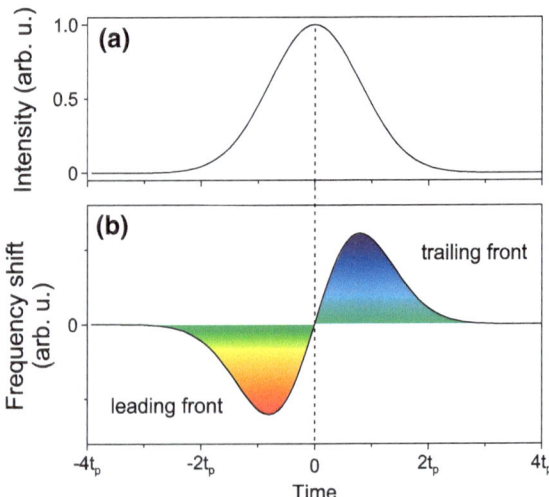

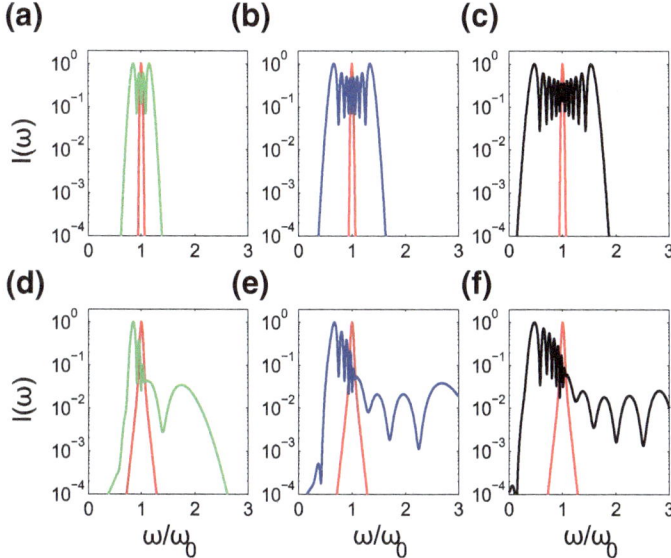

Fig. 2.3 Modulation of the spectrum resulting from Kerr-induced self-phase modulation of a laser pulse. The input pulse spectrum is shown as a red curve. Propagation distances correspond to one (green curves), two (blue curves), and three (black curves) length units, for a symmetric Gaussian pulse (a–c), or a pulse with a steeper trailing front than its leading front

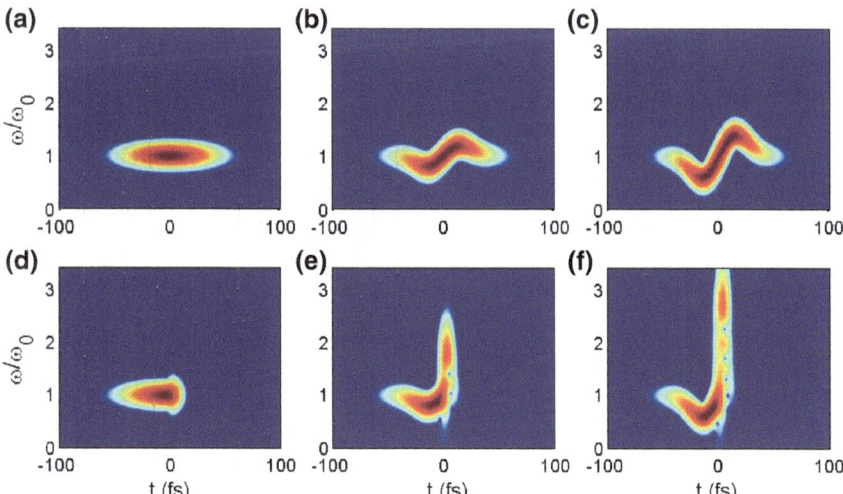

Fig. 2.4 Spectrograms of a laser pulse undergoing Kerr-induced self-phase modulation. The distribution input pulses is similar to Fig. 2.3 with **a** a symmetric input Gaussian pulse and **d** a pulse with a smooth leading front and a steep trailing front. Propagation distance correspond to **b,e** one and **c,f** two unit lengths

downshift appears in the leading pulse front (negative times) and a frequency upshift in the trailing part. The steeper the trailing part, the faster the spectrum broadens toward high frequencies, Fig. 2.4e and f.

2.3   Nonlinear Absorption and Ionization

The self-focusing stage is a runaway effect in the sense that as the beam self-focuses, the intensity increases and so does the nonlinear refractive index change, resulting in turn in an enhancement of the effect of self-focusing. However, in a real experiment, a beam cannot focus to a singularity with an infinitesimally small radius and an infinite intensity; the beam collapse is arrested at the nonlinear focus as saturation effects become more important when the beam intensity increases. For ultrashort laser pulses, saturation is dominated by multiphoton absorption and ionization, producing an energy loss and generating a free electron plasma, which further absorbs and defocuses the beam. The combined action of these effects limits the intensity to a certain level, which has been called the *clamping intensity* [14]. As will be detailed in Sect. 2.5, the clamping intensity depends on the material through the order of multiphoton absorption: $K = \langle U_g/\hbar\omega_0 \rangle + 1$, where U_g is the bandgap, and $\hbar\omega_0$ is the photon energy [15, 16]. The higher is the order of multiphoton absorption, the higher is the clamping intensity and the smaller is the limiting beam diameter at the nonlinear focus, so the larger spectral broadening is produced.

These simple considerations provide a plausible explanation of the experimentally observed bandgap dependence of the supercontinuum spectral extent and suggest that the broadest supercontinuum spectra could be attained in wide bandgap dielectrics [15, 16], provided the high clamping intensity can be maintained over a long distance. In contrast, self-focusing in the case of low multiphoton transitions at low order ($K < 3$) cannot produce supercontinuum. However, it is interesting to note the inverse relationship between the bandgap U_g and the nonlinear refractive index coefficient n_2; the larger is the bandgap, the smaller is the value of n_2 [17]. This is quite a paradox, since n_2 defines the strength of self-focusing and self-phase modulation, which are the fundamental physical effects behind femtosecond filamentation.

2.4   Plasma Effects

2.4.1   Transition of Electrons from the Valence to the Conduction Band

Ionization of a dielectric medium requires transitions of electrons from the valence band to the conduction band, and therefore the absorption of a minimum number K of photons corresponding to an energy larger than the bandgap. In the multiphoton regime, the transition rate is proportional to I^K. In addition to the energy lost by the pulse, this process results in conduction electrons whose density ρ is determined by the rate equation

$$\frac{\partial \rho}{\partial t} = \frac{\beta_K}{K \hbar \omega_0} I^K, \tag{2.11}$$

where β_K is the multiphoton absorption (ionization) coefficient.

2.4.2 Refractive Index Change

The conduction electrons, in turn, modify the permittivity of the medium. The refractive index becomes the sum of the refractive index of the unperturbed sample and the response of the conduction electrons, seen as a free electron gas

$$n = n_0 - \frac{\rho}{2 n_0 \rho_c} \tag{2.12}$$

where $\rho_c = \epsilon_0 m_e \omega_0^2 / e^2$ denotes the critical plasma density beyond which the plasma becomes opaque to an electromagnetic radiation of frequency ω_0. Here m_e and e denote the electron mass and charge, respectively.

For a Gaussian beam, the plasma density is typically larger in the center of the beam compared to the feet. The change of refractive index is therefore negative in the center of the beam. This means that the electrons in the conduction band will induce a phase curvature similar to that of a beam passing a defocusing lens. Upon further propagation, the beam intensity will decrease and its diameter will increase. For these reasons, the effect is called plasma defocusing.

2.4.3 Plasma-Induced Phase Modulation

Associated with plasma defocusing, which relies on the spatial dependence of the plasma density profile, another effect similar to self-phase modulation is induced by the fact that the plasma density also depends on time. This imparts a nonlinear phase change

$$\phi_{nl}(t) = -\frac{\omega_0 L}{c} \frac{\rho(t)}{2 n_0 \rho_c}, \tag{2.13}$$

where the dependence of the electron density upon the propagation distance has been neglected. Following the derivation performed for the optical Kerr effect, we can express the variation of the instantaneous frequency for a Gaussian laser pulse of duration t_p by using the photoionization rate (2.11)

$$\delta\omega(t) = \frac{\omega_0 L}{c} \frac{\beta_K I_0^K}{K\hbar\omega_0 2 n_0 \rho_c} \exp\left(-\frac{2Kt^2}{t_p^2}\right). \tag{2.14}$$

The effect, called plasma-induced phase-modulation, gives rise to new frequencies which are all positive, hence corresponding to a blueshift of the spectrum.

Figure 2.5 shows holographic measurements and reconstruction of the amplitude and phase contrast at the same propagation distance for a self-focusing pump beam in sapphire, with input pulse energies varying from 1.2 µJ (1.2 P_{cr}, where $P_{cr} \sim 3.3$ MW) to 14 µJ. The phase contrast provides information on the refractive index change experienced by the pulse while the amplitude contrast indicates a change in transmission. For the lowest pulse energy, the negative phase contrast originates from Kerr-induced self-phase modulation and is the only detected effect. At this propagation distance, no plasma is detected in the amplitude contrast image. For the input energy of 2.4 µJ, the phase contrast indicates a negative phase shift at the location of the leading pulse front and a positive phase shift corresponding to the effect of the plasma generated in the trailing part of the pulse. This is corroborated by the amplitude contrast image clearly showing the plasma trail. Similar effects of absorption and self-phase modulation were recorded by further increasing the pulse energy. In the wake of the pulse, the length of the plasma tail increases with the pulse energy. The maximum Kerr-induced negative phase shift of -0.15 rad is limited by

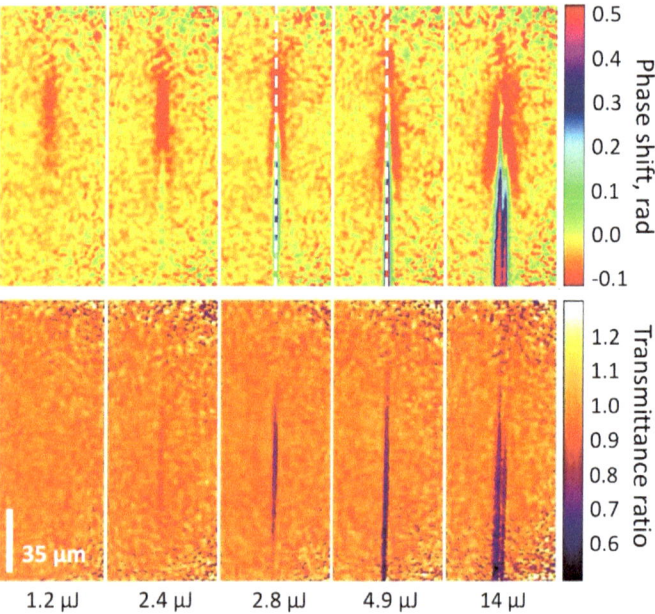

Fig. 2.5 Holographic images of a single light filament in sapphire for different input pulse energies. The top (bottom) row shows phase (amplitude) contrast (respectively). The pulse propagates from bottom to top. Reproduced from [18] by permission from the Optical Society of America

intensity clamping for increasing pulse energies up to 4.9 µJ and by the onset of multiple filamentation above this energy.

2.4.4 The Drude–Lorentz Model

The Drude model was originally derived to explain the conductivity of metals. The model assumes that electrons are delocalized and can be viewed as a *free electron gas* within the metal formed by a network of positively charged ions. Long-range interactions between electrons and ions or between the electrons are neglected. An electron is assumed to interact with its environment is via instantaneous collisions, with a characteristic collision frequency v_c which is independent of the nature of collisional processes. An electron subject to the laser pulse undergoes the action of the Lorentz force, thus, the average velocity of the electron *fluid* is governed by Newton's equation

$$\frac{d\mathbf{v}}{dt} = -\frac{e}{m}\mathbf{E} - v_c\mathbf{v} \tag{2.15}$$

The conduction current $\mathbf{j} = -\rho e\mathbf{v}$ can therefore be expressed, assuming complex notations, as

$$\mathbf{j} = \frac{e^2\rho}{m}\frac{\mathbf{E}}{v_c - i\omega_0} = \frac{\rho}{\rho_c}\frac{\omega_0^2(v_c + i\omega_0)}{v_c^2 + \omega_0^2}\epsilon_0\mathbf{E} \tag{2.16}$$

where ω_0 denotes the frequency of the laser pulse. From the formal analogy between the current and the time derivative of the polarization density $\mathbf{j} \equiv d\mathbf{P}/dt = -i\omega_0\epsilon_0\chi\mathbf{E}$, the susceptibility corresponding to plasma effects is obtained and reads

$$\chi = -\frac{\rho}{\rho_c}\frac{1 - iv_v/\omega_0}{1 + v_c^2/\omega_0^2}. \tag{2.17}$$

The real part of the susceptibility is associated with a change of refractive index that extends expression (2.12) to the case where collisional processes are taken into account:

$$\delta n = n - n_0 = -\frac{\rho}{2n_0\rho_c(1 + v_c^2/\omega_0^2)}, \tag{2.18}$$

This change of refractive index is responsible for beam defocusing due to the plasma. The imaginary part of the susceptibility is responsible for absorption of energy due to the plasma, simply called plasma absorption.

For bulk transparent materials, the Drude–Lorentz model is widely applied in the literature, assuming collisional processes of electrons in the conduction band are described by a single effective collision time.

2.5 Intensity Clamping

Several physical effects can arrest self-focusing and beam collapse. As already mentioned, the combined action of nonlinear absorption of energy and plasma defocusing dominates saturation mechanisms. However, these effects are fundamentally different in nature since nonlinear absorption acts directly on the field amplitude while the electron plasma acts essentially on the field phase.

Nonlinear absorption of energy leads to a reduction of high intensities, and therefore, to a decrease of power in the most intense central part of a pulse. A lower power means a lower effect of self-focusing while a lower intensity means a lower nonlinear refractive index change $n_2 I$, accompanied by a lower effect of self-phase modulation, thereby stopping the generation of new frequency components in the supercontinuum.

Plasma defocusing is an effect mediated by the spatial phase (beam curvature) induced by the change of refractive index localized in the central region of the beam. Since self-focusing is also an effect mediated by the beam curvature, it was proposed that the clamping intensity in a filament corresponds to the balance of refractive index changes [14]:

$$n_2 I_{\mathrm{cl}} = \frac{\rho_{\mathrm{cl}}}{2 n_0 \rho_c}, \tag{2.19}$$

where I_{cl} and ρ_{cl} denote the values for the clamping intensity and electron density, respectively. If the plasma density is roughly evaluated as $\rho = \beta_K (K\hbar\omega_0)^{-1} I^K t_p$, by assuming a top-hat pulse of duration t_p, and introduced into Eq. (2.19), an expression for the clamping intensity is obtained as

$$I_{\mathrm{cl}} = \left(\frac{2 n_0 n_2 \rho_c K \hbar\omega_0}{\beta_K t_p} \right)^{1/(K-1)}. \tag{2.20}$$

Equation (2.19) is nothing but a simplification of the similar balance equation, proposed by Braun et al. [7], that considers diffraction as a competing defocusing effect

$$n_2 I_{\mathrm{cl}}^* = \frac{\rho_{\mathrm{cl}}^*}{2 n_0 \rho_c} + \frac{(1.22\lambda_0)^2}{8\pi n_0 w_0^2}, \tag{2.21}$$

where w_0 denotes the width of the filament. Solving Eq. (2.21) for the intensity requires an additional assumption linking the intensity to the beam width: it is assumed that the filament is a self-trapped mode with *e.g.*, a Gaussian beam shape, carrying exactly one critical power for self-focusing, and thus, the relation $I_{\mathrm{cl}}^* = 2 P_{cr}^* / \pi w_0^2$ can be introduced in Eq. (2.21), which admits a solution

$$I_{\mathrm{cl}}^* = \left(\frac{0.76 n_0 n_2 \rho_c K \hbar\omega_0}{\beta_K t_p} \right)^{1/(K-1)} = 0.38^{1/(K-1)} I_{\mathrm{cl}}. \tag{2.22}$$

The clamping intensity therefore depends on the material properties via the non-linear index coefficient n_2, the multiphoton ionization coefficient β_K and the number of photons K involved in the process. In condensed media, the maximum clamped intensities up to tens of TW/cm^2 inside the nonlinear medium are estimated from Eqs. (2.20) and (2.22) at near-infrared wavelengths. Good agreement is obtained with the fact that materials with wide bandgap exhibit a higher clamping intensity. However, these models only account for the balance between self-focusing and defocusing by free electron plasma generated by multiphoton absorption.

More generally, the assumptions used to derive Eqs. (2.20) and (2.22) provide limitations for the validity of the estimations of the clamping intensity. It is interesting to discuss these limitations:

- A self-trapped mode (with Gaussian beam shape) was assumed while numerical simulations of femtosecond filamentation show a dynamic process [19].
- Intensity clamping was assumed to be solely determined by effects acting on the beam curvature while there are key players such as multiphoton absorption directly modifying the intensity, or chromatic dispersion leading to a redistribution of power within the pulse.
- The pulse structure was neglected. Plasma defocusing, which acts mainly on the trailing edge of the pulse where the electron plasma density has accumulated as the pulse intensity increased, was assumed to efficiently compete with instantaneous self-focusing which acts on the central part of the pulse.
- If the input beam energy and focusing conditions are properly chosen, the catastrophic avalanche ionization does not come into play, so the plasma density is kept below the critical value (10^{21} cm^{-3}) and optical damage of the material is avoided. However, Eqs. (2.20) and (2.22) are not valid for tight focusing geometries, since the effect of plasma defocusing cannot efficiently compete with linear focusing [20], leading to the possible occurrence of material breakdown.
- A constant pulse duration was assumed. However, plasma defocusing and absorption become relevant for the input pulse intensities of few TW/cm^2, contributing to significant shortening of the pulses before the nonlinear focus.

Despite the extrapolations leading to expressions (2.20) and (2.22), the intensity clamping effect was verified experimentally by increasing the laser pulse energy and measuring a physical quantity that depends on the pulse intensity in the filament. A saturation of the measured signal is interpreted as a signature of intensity clamping. For example, Liu et al. have observed saturation of the extent of the supercontinuum spectrum [21]. They recorded a constant width of the supercontinuum spectrum for a wide range of pulse energies above a threshold input laser energy for supercontinuum generation. They proposed that intensity clamping is responsible for limited blueshift and redshift associated with Kerr and plasma-induced self-phase modulations, the main actors in the generation of new frequencies in the supercontinuum. Indeed from Eqs. (2.10) and (2.14), a limited value of intensity will lead to bounds for the extrema of newly generated frequencies.

This interpretation, however, implicitly relies on several assumptions in the derivation of Eqs. (2.10) and (2.14). In particular, competing mechanisms in the limitation of the spectral extent of the supercontinuum were neglected. The spectral extent of the supercontinuum is an integrated diagnostic, not solely depending on the highest intensity reached during laser–matter interaction but also on the length of the interaction medium. For this reason, it cannot provide a sufficient clue for the physical mechanism or for the accurate level of intensity clamping. Competing mechanisms in the saturation of intensity itself were also neglected such as, e.g., nonlinear absorption of energy and the formation of multiple filaments, which redistributes the pulse energy over multiple hot spots.

2.6 Chromatic Dispersion

Chromatic dispersion reflects the fact that waves of different frequencies travel at different velocities in a dielectric medium. A short pulse with a spectral extent of a few to a few tens of nanometers will therefore undergo chromatic dispersion, manifesting itself in a pulse temporal spreading. This results from the fact that the different frequencies of the wavepacket separate from each other. The fastest frequencies will eventually propagate in the leading edge of the pulse, similarly to the fastest runners in a group. As it propagates, the pulse develops a chirp, i.e., a variation of its instantaneous frequency with time. Chromatic dispersion can also manifest itself in temporal compression if the fastest frequencies are initially in the trailing edge of the pulse, i.e., if the pulse carries a chirp when its propagation starts. Similarly to the case where the fastest runners in a group start behind and catch up the group, the fastest frequencies will get closer to the slowest ones, resulting in a shorter pulse.

In a material, chromatic dispersion is characterized by the dependence of the refractive index $n(\omega)$ as a function of frequency. The knowledge of this quantity requires measurements with multiple light sources. For most materials, measurements for the refractive index in their transparency range can be fitted by Sellmeier dispersion relations

$$n^2(\lambda) = a_0 + \sum_{j=0} N \frac{a_j}{\lambda_j^2 - \lambda^2}. \tag{2.23}$$

An extended database can be found in print [22] and online [23]. It is convenient to classify chromatic dispersion into the normal and the anomalous range. In this aim, the propagation constant, expressed as

$$k(\omega) = n(\omega)\frac{\omega}{c}, \tag{2.24}$$

is expanded around the central frequency of the laser pulse considered for propagation in the medium

$$k(\omega) = k(\omega_0) + k_0'(\omega - \omega_0) + \frac{k_0''}{2}(\omega - \omega_0)^2, \tag{2.25}$$

where $k_0' = \partial k / \partial \omega |_{\omega_0}$ denotes the inverse of the group velocity and $k_0'' = \partial^2 k / \partial \omega^2 |_{\omega_0}$ denotes the group velocity dispersion (GVD) coefficient. For a given material, the region of normal GVD corresponds to frequencies with a positive GVD coefficient ($k_0'' > 0$); the low (red-shifted) frequencies of the spectrum travel faster than the high (blue-shifted) ones. The opposite is true in the region of anomalous GVD ($k_0'' < 0$), where high (blue-shifted) frequencies are faster.

Numerical studies uncovered that besides the intensity clamping, the chromatic dispersion is an equally important player, which determines the extent and shape of the supercontinuum spectrum [24, 25]. The role of chromatic dispersion could be fairly evaluated in the framework of the effective three-wave mixing, which interprets supercontinuum generation as the emergence of new frequency components due to scattering of the incident optical field from the material perturbation via nonlinear polarization [24, 25]. From a simple and practical viewpoint, this approach suggests that lower chromatic dispersion allows fulfillment of the phase matching condition for a broader range of scattered spectral components, that is, supports larger spectral broadening and vice versa.

The sign of GVD basically defines the emerging temporal dynamics of femtosecond filament and so the temporal and spectral content of the SC, see e.g., [26] for an illustrative numerical study. Figure 2.6 compares the numerically simulated temporal evolution of femtosecond filament and respective spectral dynamics in sapphire crystal, in the ranges of normal (the input wavelength 800 nm), zero (1.3 μm), and anomalous (2.0 μm) GVD.

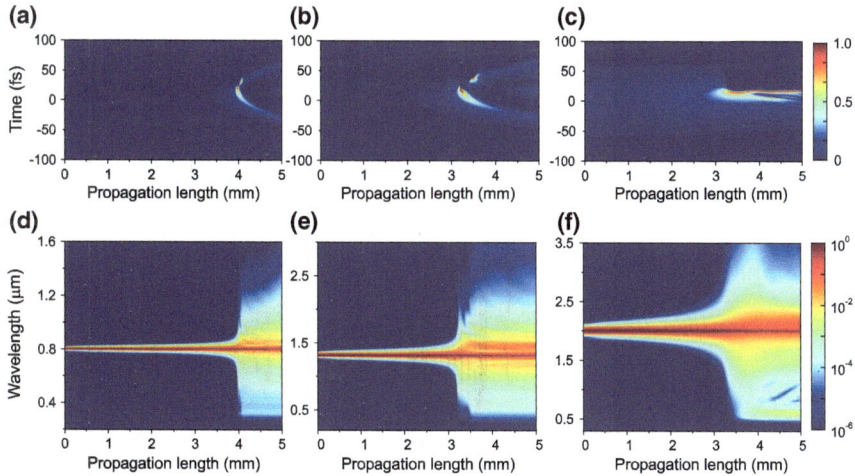

Fig. 2.6 Top row: numerically simulated temporal dynamics of 100 fs laser pulses propagating in sapphire crystal with the input wavelengths of **a** 800 nm, **b** 1.3 μm, **c** 2.0 μm, representing the filamentation regimes of normal, zero and anomalous GVD, respectively. Bottom row shows the corresponding spectral dynamics. Notice, how the spectral broadening in the regimes of normal and zero GVD is associated with the pulse splitting, and the spectral broadening in the regime of anomalous GVD is associated with pulse self-compression

2.7   Self-steepening and Space-Time Focusing

The interplay between the nonlinear effects and chromatic dispersion gives rise to self-steepening. The Kerr nonlinearity modifies not only the refractive index of the medium as indicated in Eq. (2.3) but also the group index, via the interplay with dispersion, as expressed by the total group index

$$n^{(g)} = n_0^{(g)} + n_2^{(g)}, \tag{2.26}$$

where $n_0^{(g)} \equiv n(\omega_0) + \omega_0 dn/d\omega|_{\omega_0}$ and $n_2^{(g)}$ is proportional to n_2 and positive. As a result, the intense part of the pulse therefore travels at a smaller velocity with respect to the low intensity feet, resulting in a steepening of the trailing edge of the pulse [27]. An optical shock wave (infinitely steep trailing edge) can even form for few-cycle pulses propagation in highly nonlinear Kerr media.

Space-time focusing refers to the effect of diffraction into different cone angles for the different frequency components of an ultrashort pulse. This effect is not only significant for few-cycle pulses, but also for multi-cycle pulses as a mediator of space-time couplings through propagation.

2.8   Four-Wave Mixing and Phase Matching

Four-wave mixing (FWM) is a parametric nonlinear process arising in Kerr media. It denotes the interaction between two or three photons (frequencies) producing two or one new photons (frequencies). Four-wave mixing is usually expressed as a conservation equation for the energy of the incoming photons

$$\omega_1 + \omega_2 = \omega_3 + \omega_4, \tag{2.27}$$

with the implicit assumption that the frequencies ω_i can be positive or negative.

Four-wave mixing is a phase-sensitive process, i.e., the efficiency of the process is strongly affected by the relative phases of interacting fields. If the relative phases can be maintained, the effect can accumulate over long propagation distances and a signal of new frequency can be efficiently generated. This requires *phase matching conditions* to be satisfied.

A particular case of four-wave mixing is the third harmonic generation. The optical Kerr effect gives rise to a nonlinear polarization with a component at frequency $3\omega_0$ when the frequency of the applied field is ω_0. This component of the nonlinear polarization is responsible for third harmonic generation. It can be viewed as a four-wave mixing process in which three photons at frequency ω_0 are annihilated to create one photon of frequency $3\omega_0$.

Four-wave mixing is related to self-phase modulation and cross-phase modulation: all these effects originate from the same (Kerr) nonlinearity and differ only in terms of degeneracy of the waves involved. Cross-phase modulation denotes the effect on the phase of a monochromatic beam at a given wavelength resulting from its interaction with a beam at another wavelength through the optical Kerr effect. This effect arises due to the dependence of the refractive index of the medium on the intensity of a beam. Cross-phase modulation is the impact of the refractive index on another beam at a different wavelength.

References

1. Dudley, J.M., Genty, G., Coen, S.: Supercontinuum generation in photonic crystal fiber. Rev. Mod. Phys. **78**, 1135–1184 (2006)
2. Alfano, R.R., Shapiro, L.: Emission in the region 4000 to 7000 Å via four photon coupling in glass. Phys. Rev. Lett. **24**, 584–587 (1970)
3. Alfano, R.R., Shapiro, L.: Observation of self-phase modulation and small-scale filaments in crystals and glasses. Phys. Rev. Lett. **24**, 592–594 (1970)
4. Chin, S.L., Hosseini, S.A., Liu, W., Luo, Q., Thberge, F., Aközbek, N., Becker, A., Kandidov, V.P., Kosareva, O.G., Schroeder, H.: The propagation of powerful femtosecond laser pulses in optical media: physics, applications, and new challenges. Can. J. Phys. **83**, 863–905 (2005)
5. Couairon, A., Mysyrowicz, A.: Femtoseconmd filamentation in transparent media. Phys. Rep. **441**, 47–190 (2007)
6. Kandidov, V.P., Shlenov, S.A., Kosareva, O.G.: Filamentation of high-power femtosecond laser radiation. Quantum Electron. **39**, 205–228 (2009)
7. Braun, A., Korn, G., Liu, X., Du, D., Squier, J., Mourou, G.: Self-channeling of high-peak-power femtosecond laser pulses in air. Opt. Lett. **20**, 73–75 (1995)
8. Strickland, D., Mourou, G.: Compression of amplified chirped optical pulses. Opt. Commun. **56**, 219–221 (1985)
9. Chiao, R.Y., Garmire, E., Townes, C.H.: Self-trapping of optical beams. Phys. Rev. Lett. **13**, 479–482 (1964)
10. Porras, M.A., Parola, A., Faccio, D., Couairon, A., Di Trapani, P.: Light-filament dynamics and the spatiotemporal instability of the townes profile. Phys. Rev. A **76**, 011803(R) (2007)
11. Marburger, J.H.: Self-focusing: theory. Prog. Quantum Electron. **4**, 35–110 (1975)
12. Dawes, E.L., Marburger, J.H.: Computer studies in self-focusing. Phys. Rev. **179**, 862–868 (1969)
13. Dubietis, A., Couairon, A., Kučinskas, E., Tamošauskas, G., Gaižauskas, E., Faccio, D., Di Trapani, P.: Measurement and calculation of nonlinear absorption associated with femtosecond filaments in water. Appl. Phys. B **84**, 439–446 (2006)
14. Kasparian, J., Sauerbrey, R., Chin, S.L.: The critical laser intensity of self-guided light filaments in air. Appl. Phys. B **71**, 877–879 (2000)
15. Brodeur, A., Chin, S.L.: Band-gap dependence of the ultrafast white-light continuum. Phys. Rev. Lett. **80**, 4406–4409 (1998)
16. Brodeur, A., Chin, S.L.: Ultrafast white-light continuum generation and self-focusing in transparent condensed media. J. Opt. Soc. Am. B **16**, 637–650 (1999)
17. Sheik-Bahae, M., Hagan, D.J., Van Stryland, E.W.: Dispersion and band-gap scaling of the electronic Kerr effect in solids associated with two-photon absorption. Phys. Rev. Lett. **65**, 96–99 (1990)
18. Šiaulys, N., Melninkaitis, A., Dubietis, A.: In situ study of two interacting femtosecond filaments in sapphire. Opt. Lett. **40**, 2285–2288 (2015)
19. Mlejnek, M., Wright, E.M., Moloney, J.V.: Dynamic spatial replenishment of femtosecond pulses propagating in air. Opt. Lett. **23**, 382–384 (1998)

20. Kiran, P.P., Bagchi, S., Arnold, C.L., Krishnan, S.R., Kumar, G.R., Couairon, A.: Filamentation without intensity clamping. Opt. Express **18**, 21504–21510 (2010)
21. Liu, W., Petit, S., Becker, A., Aközbek, N., Bowden, C.M., Chin, S.L.: Intensity clamping of a femtosecond laser pulse in condensed matter. Opt. Commun. **202**, 189–197 (2002)
22. Weber, M.J.: Handbook of Optical Materials. CRC Press, London (2003)
23. https://refractiveindex.info/
24. Kolesik, M., Katona, G., Moloney, J.V., Wright, E.M.: Physical factors limiting the spectral extent and band gap dependence of supercontinuum generation. Phys. Rev. Lett. **91**, 043905 (2003)
25. Kolesik, M., Katona, G., Moloney, J.V., Wright, E.M.: Theory and simulation of supercontinuum generation in transparent bulk media. Appl. Phys. B **77**, 185–195 (2003)
26. Kolesik, M., Wright, E.M., Moloney, J.V.: Interpretation of the spectrally resolved far field of femtosecond pulses propagating in bulk nonlinear dispersive media. Opt. Express **13**, 10729–10741 (2005)
27. DeMartini, F., Townes, C.H., Gustafson, T.K., Kelley, P.L.: Self-steepening of light pulses. Phys. Rev. **164**, 312–323 (1967)

Chapter 3
Femtosecond Filamentation in Solid-State Media

Femtosecond filamentation refers to the ability of powerful femtosecond laser pulses to propagate nonlinearly over distances of several diffraction (Rayleigh) lengths in a medium with Kerr nonlinearity. The laser beam usually undergoes self-focusing and remains narrow and intense enough to deposit its energy into a needle-shaped region along the propagation axis. The strong laser-matter interaction in this region is characterized by the possible generation of an electron-hole plasma and the emission of new radiations. The phenomenon was reported to happen in air for the first time with multi-gigawatt femtosecond laser pulses in 1995 by the team of Mourou [1]. The chirped pulse amplification technology was indeed necessary before laser pulses with peak powers exceeding the critical power for self-focusing in air became routinely available in laboratories. In transparent solids, femtosecond filamentation was reported in 2001 [2]. The required peak power for self-focusing in transparent solids is only of a few megawatts and could be achieved by various modern femtosecond lasers. Manifestations of the filamentation phenomenon in solid media with longer pulses were therefore reported at the dawn of nonlinear optics, by Michael Hercher in 1964, in the form of long filamentary damage tracks caused by the propagation of an intense laser beam in glass [3].

Femtosecond filamentation has been investigated extensively during more than two decades both in gaseous and in condensed media. While numerous explanations for the rich physics involved during filamentation were and may still be debated, universal features of femtosecond filamentation were systematically recorded (see [4]), not only appearing in all types of media but also in various experimental conditions and for different lasers with central wavelengths from the ultraviolet to the near-infrared wavelength region. In the following, apart from a couple of illustrative examples taken from filamentation experiments in air, owing to their historical importance, we focus mainly on features of filamentation in transparent condensed media, i.e., solids, or liquids which are used as prototypical condensed media to avoid potential damage of the material and facilitate length tuning.

© The Author(s), under exclusive license to Springer Nature Switzerland AG 2019
A. Dubietis and A. Couairon, *Ultrafast Supercontinuum Generation in Transparent Solid-State Media*, SpringerBriefs in Physics,
https://doi.org/10.1007/978-3-030-14995-6_3

3.1 Universal Features of Femtosecond Filamentation

3.1.1 Conical Emission

The most striking manifestation of femtosecond filamentation is certainly the conical emission of white light. It is accompanied by axial supercontinuum generation. First observations of conical emission from filamentation of femtosecond laser pulses in air were reported by Nibbering et al. [5], Fig. 3.1a. The cross section of the laser beam several meters after the termination of the filament exhibits a white central spot surrounded by colored rings, whose wavelengths decrease from the central to the outer rings. Conical emission is now considered as a signature of filamentation as it was observed for filaments in media of various nature (gases, liquids, solids [6], Fig. 3.1b) and lasers of different central wavelengths. The most spectacular observations of conical emissions were performed using near-infrared or visible lasers as the newly generated colors during laser–matter interaction lie in the visible range of the electromagnetic spectrum.

3.1.2 Plasma Channel Formation

The generation of a plasma accompanying femtosecond filamentation can often be observed during experiments from the luminescence of the plasma. In air and gases, several measurement techniques confirmed the presence of a plasma in the wake of the pulse undergoing filamentation, with ionization degrees reaching 10^{-3} [7]. In

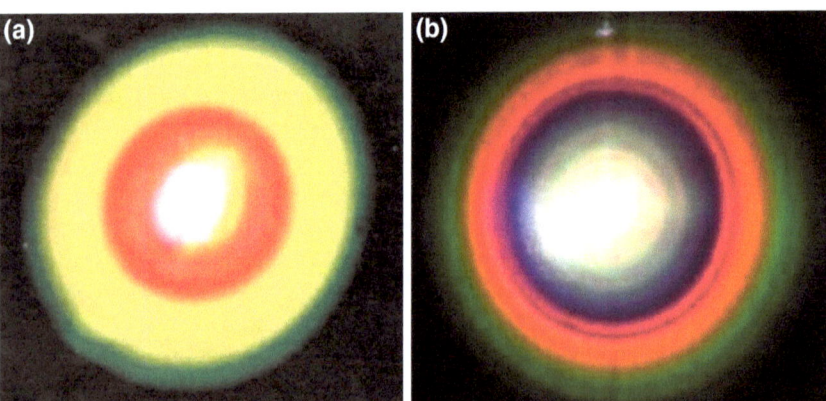

Fig. 3.1 **a** Conical emission accompanying filamentation in air. Image of the beam cross section after propagation of a terawatt laser pulse over 40 m. The diameter of the external ring is about 40 cm. Adapted from [5]. **b** Conical emission from a filament in a ZK7 glass Adapted from [6]. Reprinted by permission from the Optical Society of America

condensed media, the presence of an electron plasma is more delicate to characterize but carefully designed pump-probe experiments [8] or inline holographic techniques [9] permitted to achieve this task both in transparent solids and in liquids. The density of the plasma can reach levels up to 10^{20} cm^{-3}, depending on the focusing geometry [10]. It is the detection of a plasma in the wake an intense pulse undergoing filamentation in air that led to the proposal that filamentation results from an interplay between Kerr self-focusing and defocusing due to the self-generated electron plasma [1]. This explanation is still extremely popular, though it falls short in explaining the filament ability to self-heal beyond an obstacle.

3.1.3 Filament Robustness and Energy Reservoir

Filaments are robust. A narrow filament can be arrested by an obstacle and be regenerated by the surrounding beam beyond the obstacle. This property was shown by several teams for filaments in air and in condensed media [11–16]. This robustness was explained by the concept of energy reservoir: For a beam carrying several powers for self-focusing, the intense peak or filamentary part carries only a fraction of the total beam power, evaluated to approximately one critical power for self-focusing. A droplet of rain, an obstacle, a stopper, is tantamount to a sudden nonlinear absorption of energy that corresponds to P_{cr}. Similarly to the Arago spot experiment, the filament was observed to self-heal, after a distance of a Rayleigh length, i.e., a new intense and narrow peak continues to propagate almost as if the obstacle had been absent. The surroundings of the central peak can pass the obstacle freely and naturally provide the energy for the reconstruction of the central peak beyond the obstacle. For this reason, the low intensity part of the beam surrounding the filament is called the *energy reservoir* [11].

The concept of energy reservoir put into question the explanation of filamentation as a self-guided mode. In a self-guiding process, nonlinear propagation over several Rayleigh lengths requires self-focusing to at least balance diffraction, or overcome it if other defocusing mechanisms play a role, such as the refraction due to the self-generated plasma. The original experiments for femtosecond filamentation in air were interpreted in terms of an extension of the concept of self-trapping supported by the Townes mode that relies on a balance between diffraction and self-focusing [17]. Plasma defocusing was supposed to stabilize and guide the nonlinear beam [1]. The power content for the resulting extended Townes mode or spatial soliton was therefore, not astonishingly, quantified to a power close to that of the Townes mode, i.e., P_{cr}. However, complimentary experiments were realized to establish the relative roles of the energy reservoir and self-guiding mechanism in sustaining the long-distance propagation of the filament [12, 13]. The model of a filament sustained by a balance between Kerr self-focusing, plasma defocusing and diffraction was challenged by blocking the energy reservoir by means of a pinhole with diameter only slightly larger than that of the intense core. This prevented the energy reservoir the to play a role while letting the filament propagate freely through the pinhole. The result was that the filament disappeared roughly one diffraction length after the pinhole. Hence, the energy reservoir plays a key role in sustaining filamentation over long distances. The universal feature of long-distance propagation of the intense core of filaments does not proceed from a balance between focusing and defocusing effects.

3.1.4 Conical Waves

A natural framework that consistently explains the physics of the energy reservoir is Bessel beam propagation. A Bessel beam is a propagation-invariant (linear) solution to the Maxwell equations. It is a monochromatic beam with radial intensity distribution described by a Bessel function, corresponding to a scalar field

$$E(r, z) = E_0 \exp(ik_z z) J_0(k_0 \sin \theta r)$$

where J_0 is the lowest order Bessel function and k_0 the wavenumber of the monochromatic beam with longitudinal and transverse components $k_z = k_0 \cos \theta$ and $k_\perp = k_0 \sin \theta$, respectively. A distinguishing feature of Bessel beams is that the energy flow is not directed along the z-propagation axis as in conventional Gaussian beams. Bessel beams can be viewed as a superposition of plane waves whose wavevector lies on a cone-shaped surface around the propagation axis, with cone half-angle θ, see Fig. 3.2a. The main lobe of the Bessel beam thus appears as a very intense and localized interference peak with dimensions of the order of a few wavelengths, which propagates without spreading. This peak is surrounded by slowly decaying tails in the form of concentric rings that contain the major part of the beam energy. In this sense, the Bessel beam is a perfect illustration of the concept of energy reservoir surrounding a filament.

In contrast with the ideal Bessel beam which has infinite energy, experimental realizations of Bessel beams carry a finite energy due to the fact that real optical systems have finite transverse dimensions. For instance, a Bessel–Gauss beam is generated by focusing a Gaussian beam with an axicon. In result, the Bessel– Gauss beam is quasi-propagation invariant, i.e., it does not spread over a limited distance called the Bessel zone, that is anyway orders of magnitude larger than the diffraction length of the intense peak, Fig. 3.2a.

The idea of monochromatic propagation invariant beams has been generalized to the polychromatic case by Saari and Reivelt [18] in optics, building on works in acoustics by Lu and Greenleaf [19]. They demonstrated that a linear superposition of monochromatic Bessel beams having different conical angles at different frequencies constitutes a linear propagation invariant solution to the Maxwell equations

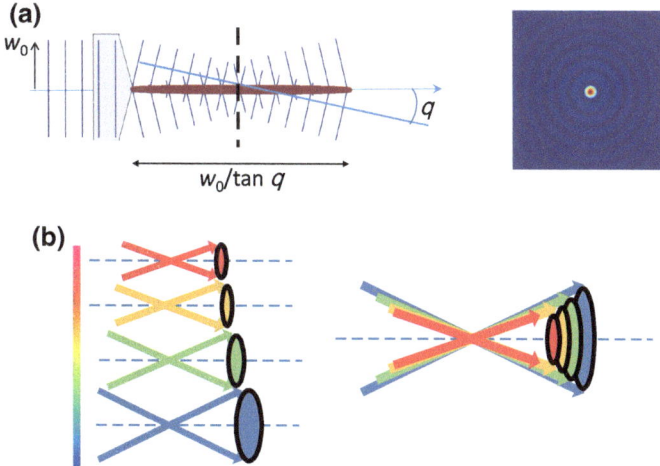

Fig. 3.2 **a** Schematic representation of a Bessel beam generated by focusing a Gaussian beam of waist w_0 with an axicon. Wavevectors lie on a cone-shaped surface of cone half-angle θ and the intensity distribution follows the Bessel function within the Bessel zone of length $w_0/\tan\theta$. **b** Schematic representation of a polychromatic conical wave as a superposition of Bessel beams having different wavelengths

(Fig. 3.2b). An expression for these solutions in a dispersive medium of refractive index $n(\omega)$ takes the form:

$$E(r, t) = \int_{-\infty}^{+\infty} S(\omega) J_0(k_\perp r) \exp(ik_z z - \omega t)d\omega,$$

where $S(\omega)$ is any spectral function and the transverse wavenumber is $k_\perp = k(\omega)\sin\theta = (n(\omega)\omega/c)\sin\theta$. It is determined by the Bessel cone angle θ and satisfies the dispersion relation $k_\perp = \sqrt{k^2(\omega) - k_z^2}$. The longitudinal wave number $k_z = k(\omega)\cos\theta$ must be a linear function of ω in order for the wavepacket to be propagation invariant. These wavepackets do not spread in space or time and are called conical spatiotemporal waves, or X-waves in normally dispersive media, because their spatiotemporal intensity distribution possess an intense core surrounded by hyperbolic tails storing a large amount of energy. These tails are shaped as a cone (or the letter X if only one transverse dimension is considered) and this shape is the same in the near field and in the far field since it characterizes propagation invariant wavepackets. The tails play the same role as the energy reservoir for Bessel beams and filaments, i.e., they sustain the long-distance propagation of the intense peak.

In the nonlinear regime, the concept of conical waves was extended both in the monochromatic and polychromatic cases [20–22]. In the case of monochromatic beams, numerous propagation invariant nonlinear conical waves have been identified in the form of nonlinear Bessel beams that can be viewed as nonlinear eigenmodes of a nonlinear Helmholtz equation. These solutions have the specificity to be weakly localized, and therefore constitute a continuous family of solutions corresponding to the continuous spectrum of an eigenvalue problem. Nonlinear Bessel beams were obtained for two prototypical nonlinearities, namely a cubic Kerr nonlinearity and multiphoton absorption, allowing for a thorough understanding of the main features of monochromatic nonlinear conical waves [20, 21]: (i) They have the shape of a nonlinear Bessel beam with an intense core surrounded by rings storing the major part of the beam energy, playing the role of energy reservoir. (ii) Propagation invariance of these waves is due to an energy flux from the core periphery toward the intense peak, which is a sink of energy because of multiphoton absorption. (iii) Like their linear counterparts, nonlinear Bessel beams carry infinite energy; experimental realization of these waves carry finite energy and constitute quasi-invariant propagation over a finite distance, corresponding to the distance required to exhaust the energy reservoir. (iv) As nonlinear effects occur mainly in the intense core, it was shown that the diameter of the main lobe is narrower that the diameter of a linear Bessel beam with the same cone angle, due to Kerr self-focusing. The contrast in the rings is attenuated due to the energy flux induced by the energy losses [23].

In the case of polychromatic waves, propagation invariant nonlinear conical waves in dispersive media were also predicted theoretically and identified as spatiotemporal stationary solutions to unidirectional propagation models such as the Nonlinear Schrödinger equation. They are characterized by an intense core hosting nonlinear effects, and a balance of the phase modulation caused by material chromatic dispersion by means of that created by the wave angular dispersion [22, 24]. The distinguishing feature of these solutions is their conical structure, extending the concept of X-waves to the nonlinear realm. For this reason, they were called nonlinear X-waves. These conical waves were conjectured to act as possible attractors for the dynamics of nonlinear pulse propagation in dispersive media. To achieve the stationary regime, the nonlinear-induced phase modulation must also be fully compensated, which has been confirmed by numerical analysis [22] and by further experimental and numerical investigation, where the transient from the input Gaussian to output X-type object has been investigated in detail [25].

Since various nonlinear conical waves were identified theoretically as propagation invariant solutions to unidirectional propagation equations in Kerr dispersive media and observed in filamentation experiments in condensed media, striking analogies were highlighted between the properties of conical waves and ultrashort laser pulses undergoing filamentation. It was proposed that filamentation is nothing but a spontaneous transformation of the initial Gaussian pulse into a stationary wavepacket during its nonlinear propagation in a dispersive medium, i.e., into a nonlinear conical wave. This spontaneous transformation is exemplified in the results of numerical simulations shown in Fig. 3.3.

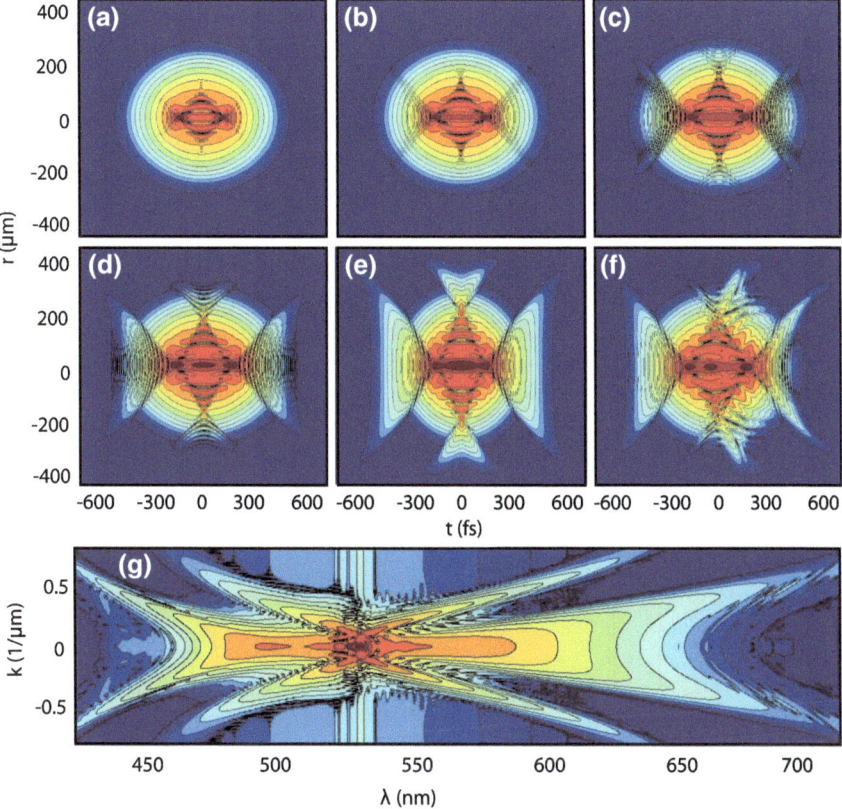

Fig. 3.3 Near-field intensity profile (shown in logarithmic scale over 6 decades) evolution of a 527 nm, 200 fs Gaussian laser pulse undergoing filamentation in water at five different distances just after the nonlinear focus at **a** 3 cm, **b** 3.2 cm, **c** 3.4 cm, **d** 3.6 cm, **e** 4 cm. **a** to **e** include only second order dispersion (k'') effects. **f** is the same as **e** but the full material dispersion and plasma generation are accounted for—see Sect. 3.2. **g** Far field ($k_\perp$, λ) spectrum corresponding to the (r, t) intensity profile shown in (**e**). Adapted from [26]

3.2 Numerical Model

Universal features of femtosecond filaments were observed in various experimental conditions and often interpreted, if not predicted, by numerical simulations of a unidirectional propagation equation coupled to a model for the material response. Interestingly, the diversity of experimental conditions is covered by this versatile model or family of models by simply modifying the parameters of the material response and input condition. The key elements of this model are presented in this section.

The earliest numerical models describing the nonlinear propagation and filamentation of ultrashort laser pulses were based on the nonlinear Schrödinger equation for the pulse envelope [27–29]:

$$\frac{\partial \mathscr{E}}{\partial z} + k_0' \frac{\partial \mathscr{E}}{\partial t} = \frac{i}{2k_0} \Delta_\perp \mathscr{E} - i \frac{k_0''}{2} \frac{\partial^2 \mathscr{E}}{\partial t^2} + ik_0 \frac{\Delta n}{n_0} \mathscr{E} - \frac{\alpha}{2} \mathscr{E}, \tag{3.1}$$

where $k_0 = k(\omega_0)$, $k_0' = dk/d\omega|_{\omega_0}$ and $k_0'' = d^2k/d\omega^2|_{\omega_0}$ are derived from $k(\omega) = n(\omega)\omega/c$, and $n(\omega)$ denotes the linear refractive index of the medium, c is the speed of light in vacuum. The term $k_0' \partial \mathscr{E}/\partial t$ in Eq. (3.1) represents unidirectional propagation of the pulse at the group velocity $k_0'^{-1}$. In practise, a Galilean change of reference frame permits to express propagation equations as a function of the local time in the pulse frame. The sign of the coefficient $k_0'' \equiv d^2k/d\omega^2|_{\omega_0}$ determines whether dispersion is normal ($k_0'' > 0$) or anomalous ($k_0'' < 0$). Note that its sign also determines the hyperbolic or elliptic nature of space and time couplings in Eq. (3.1), which are responsible for spatiotemporal reshaping into nonlinear X- and O-waves, respectively, as discussed in the previous section. The operator $\Delta_\perp$ denotes the transverse Laplacian in the plane orthogonal to the propagation direction z.

The refractive index change reads

$$\Delta n = n_2 I - \frac{\rho_e}{2n_0 \rho_c}, \tag{3.2}$$

where n_2 denotes the nonlinear index coefficient, $I \equiv \epsilon_0 c n_0 |\mathscr{E}|^2/2$, the laser pulse intensity, ρ_e the electron density, $\rho_c \equiv \epsilon_0 m \omega_0^2/e^2$ the critical plasma density beyond which the plasma becomes opaque to the laser radiation at frequency ω_0 and ϵ_0 is the permittivity of free space. Equation 3.1 is coupled with a rate equation describing the generation of free electrons by multiphoton ionization

$$\frac{\partial \rho_e}{\partial t} = R(I)(\rho_0 - \rho_e), \tag{3.3}$$

where ρ_0 is the initial density of neutral molecules and $R(I)$ is the multiphoton ionization rate of the molecules. Nonlinear losses due to ionization are accounted for by the time- dependent coefficient $\alpha = I^{-1} K \hbar \omega_0 (\partial \rho_e/\partial t)$ and K is the order of the multiphoton ionization process. Group velocity dispersion is taken into account at leading order.

Several assumptions are necessary to derive this model from Maxwell equations. (i) Eq. (3.1) is a scalar equation, assuming a linearly polarized laser pulse that remains linearly polarized along propagation. (ii) It is a paraxial propagation equation, assuming that the divergence angle of the beam is small, which implies that the beam cannot be tightly focused. (iii) Vectorial effects are neglected, consistently with the first two assumptions. (iv) Propagation is assumed to be unidirectional, preventing situations where the pulse would be reflected on, e.g., an overdense plasma, to be accurately described by this model. (v) A slowly varying envelope approximation is used, meaning that pulses with broad spectral extent around the central frequency cannot be accurately propagated. (vi) High order dispersion is neglected. (vii) Multiphoton ionization is assumed to be the dominant ionization process, limiting the validity range of the model to intensities of a few tens of TW/cm^2. (viii) The entire modeling of the medium response is phenomenological.

In spite of these simplifying assumptions, this numerical model became a widely used standard for the last decades in the field of nonlinear propagation and filamentation of intense femtosecond laser pulse in transparent dielectric media with cubic nonlinearity. This is certainly due to the fact that most of the simplifying assumptions can be easily relaxed to generalize the NLS equation into a universal class of unidirectional propagation equations. The use of the original model beyond its theoretical validity limits was hence later justified by numerical simulations of more sophisticated models that can be written in the form of a canonical propagation equation in the spectral domain for the electrical field $\hat{E} \equiv \hat{E}(\omega, k_x, k_y, z)$[30, 31]:

$$\frac{\partial \hat{E}}{\partial z} = i K(\omega, k_x, k_y)\hat{E} + i Q(\omega, k_x, k_y)\frac{\hat{P}_{NL}}{\epsilon_0}, \tag{3.4}$$

where $K(\omega, k_x, k_y) \equiv \sqrt{\omega^2 n^2(\omega)/c^2 - k_x^2 - k_y^2}$ represents the propagation constant for the modal components of the electric field, and $Q(\omega, k_x, k_y) \equiv \omega^2/2c^2 K$ (ω, k_x, k_y). The nonlinear polarization $P_{NL}(t, x, y, z)$, whose spectral representation reads as $\hat{P}_{NL}(\omega, k_x, k_y, z)$, describes the nonlinear response of the material and takes the form of constitutive relations linking $P_{NL}(t, x, y, z)$ to the electric field $E(t, x, y, z)$ including the carrier wave. Equation (3.4) encompasses all unidirectional scalar propagation models that can be derived under various approximations.

Regarding approximations: Nonparaxial diffraction is taken into account by construction. Pulses with broad spectral extension can be accurately described since Eq. (3.4) is a carrier-resolving propagation equation for the electric field $E(t, x, y, z)$. If all the approximations listed above for the NLS equation are performed, Eq. (3.4) is transformed into the NLS equation. If only a subset of approximations are performed, it is transformed into one of the numerous generalizations of Eq. (3.1). For instance, the pulse duration may be no longer than a few optical cycles and the slowly varying envelope approximation not valid, still Eq. (3.4) can be transformed into an envelope equation if the envelope is evolving slowly along the propagation direction. This less restrictive assumption is called the slowly evolving wave approximation [32]. It permits the dropping of the carrier wave $\exp(-i\omega_0 t)$, which amounts to shifting the spectrum of $\hat{E}(\omega, k_x, k_y, z)$ by a quantity ω_0 and to replacing $E(t, x, y, z)$ by the laser pulse envelope $\mathcal{E}(t, x, y, z)$ without changing the functional form of Eq. (3.4). For numerical simulations, the universal form of Eq. (3.4) is very convenient since a single algorithm is then capable of solving all propagation models that Eq. (3.4) encompasses [31]. While Equation (3.4) is still written as a scalar equation, vectorial effects can be accounted for, either by adding their contribution to the nonlinear polarization [30] or by propagating appropriate components of the Hertz vector potential [33] instead of $\tilde{E}(\omega, k_x, k_y, z)$. The dispersion landscape is described exactly in the frequency domain by means of the $k(\omega)$ relation and not by a Taylor expansion around a central frequency. As for the assumptions linked to the medium response, they can simply be relaxed by modifying the medium response.

The derivation of the NLS Eq. (3.1) from Eq. (3.4) illustrates the last two points: The paraxial approximation amounts to performing a small $(k_x^2 + k_y^2)$-expansion of K and Q as $K \sim k(\omega) - (k_x^2 + k_y^2)/2k(\omega)$ and $Q \sim \omega/2cn(\omega)$, where $k(\omega) \equiv n(\omega)\omega/c$. In this case, under this approximation, Eq. (3.4) which is still a carrier-resolving equation is transformed into the spectral representation of the forward Maxwell equation [34]. The corresponding envelope propagation equation is obtained by dropping the carrier wave [32]. The family of nonlinear Schrödinger propagation equations is obtained by expanding $k(\omega)$ as a Taylor series around the pulse carrier frequency ω_0, as

$$k(\omega) \sim k_0 + k_0'(\omega - \omega_0) + \frac{k_0''}{2}(\omega - \omega_0)^2 + \frac{k_0'''}{3!}(\omega - \omega_0)^3 + \cdots, \qquad (3.5)$$

and by truncating this expansion to the second order in $(\omega - \omega_0)$. Consistently with the paraxial approximation, K and Q are Taylor expanded as $K \sim k(\omega) - (k_x^2 + k_y^2)/2k_0$ and $Q \sim \omega_0/2cn_0$.

Limitations of the simple medium response model (3.2) and (3.3) can be overcome by generalizing expressions for the nonlinear polarization in a form that still distinguishes the responses of bound and free electrons:

$$P_{NL} = P_{bound} + P_{free}. \qquad (3.6)$$

The response of bound electrons includes two contributions to the Kerr effect: A quasi-instantaneous electronic response and a delayed Raman response due to rovibrational modes of the lattice. Assuming that the total cubic susceptibility $\chi^{(3)}$ is constant within the frequency range of interest, the nonlinear polarization induced by bound electrons is expressed as a weighted sum of the delayed and instantaneous contributions to the Kerr effect with susceptibility fractions α and $1 - \alpha$, respectively:

$$P_{bound} = \epsilon_0 \chi^{(3)} \left(\int_{-\infty}^{+\infty} R(t - t') E^2(t') dt' \right) E(t), \qquad (3.7)$$

$$R(t) = (1 - \alpha)\delta(t) + \alpha H(t) \Omega \exp(-\Gamma t) \sin(\Lambda t), \qquad (3.8)$$

where $\delta(t)$ is the Dirac delta function, $H(t)$ is the Heaviside step function, and $\Omega = \frac{\Lambda^2 + \Gamma^2}{\Lambda}$, with Γ and Λ being the characteristic frequencies for the Raman response of the dielectric medium. The nonlinear polarization P_{bound}, is responsible for two of the most relevant effects in filamentation and supercontinuum generation: self-focusing and self-phase modulation (discussed in the previous chapter). For a carrier-resolving model, this term also accounts for third harmonic generation and generation of other low order odd harmonics by cascaded four-wave mixing. For an envelope propagation model designed to simulate the supercontinuum generation over a limited spectral region, if no spectral overlap with the third harmonic is expected, it is sufficient to replace the field squared $E^2(t)$ by the squared modulus of the complex envelope $|\mathscr{E}|^2$ in the nonlinear response.

The model for free electron generation can also be generalized since the propagation model only requires a constitutive relation for P_{free}. Keeping the framework of a rate equation describing the evolution of the plasma density ρ_e, the effects of optical field ionization, avalanche ionization and plasma recombination can be modeled as

$$\partial_t \rho_e = W(I)(\rho_{\text{nt}} - \rho_e) + \frac{\sigma}{U_g}\rho_e I + \partial_t \rho_e|_{\text{rec}}, \qquad (3.9)$$

where $W(I)$ denotes the intensity dependent photo-ionization rate, ρ_{nt} is the neutral density in the valence band, U_g is the energy gap between the valence and the conduction band and σ is the cross section for inverse Bremsstrahlung (avalanche rate $\sigma I/U_g$). The response of free electrons is conveniently described by a current, acting as a source term in the propagation equation. The total current is linked to the nonlinear polarization induced by free electrons, P_{free}, and is contributed by two components:

$$\partial_t P_{\text{free}} = J_e + J_{\text{loss}}, \qquad (3.10)$$

$$\partial_t J_e + v_c J_e = \frac{e^2}{m}\rho_e E, \qquad (3.11)$$

$$J_{\text{loss}} = \epsilon_0 c n_0 \frac{W(I)}{I} U_g(\rho_{\text{nt}} - \rho_e)E. \qquad (3.12)$$

Equation (3.11) is based on the Drude model and describes the motion of electrons accelerated by the laser field, undergoing friction at a rate of v_c due to collisions with ions. J_{loss} is responsible for the loss of energy necessary to ionize the medium and is described by the phenomenological equation Eq. (3.12). As a source term in the propagation Eq. (3.4), J_e is responsible for plasma defocusing and plasma absorption.

Illustrative results presented below or reviewed throughout this book encompass a wide range of realistic experimental scenarios supported by numerical simulations based on the above general propagation and laser–matter interaction model. Pulse and material parameters are adapted case by case to each experimental situation. For further details about the theoretical background, basic building blocks and tools to perform numerical simulation with proper understanding of the underlying physical effects, the reader is referred to Ref. [31]. A classification of various approaches to optical field evolution equations is also provided in Ref. [35], together with a presentation of light–matter interaction models and methods that can be integrated with time- and space-resolved simulations.

3.3 Supercontinuum Generation Under Normal GVD

Supercontinuum generation in the normal GVD region of wide bandgap dielectric materials has received the largest theoretical and experimental attention. This is not surprising since this spectral range spanning UV to near IR wavelengths is readily accessible by modern femtosecond solid-state lasers and their harmonics.

In the region of normal GVD of dielectric media, the self-focusing dynamics and spectral broadening at and beyond the nonlinear focus is accompanied by pulse splitting phenomena. Using propagation models of different complexity, pulse splitting in dielectric media with normal dispersion was foreseen theoretically as a mechanism which contributes to arresting the collapse of ultrashort pulses with input power just slightly above P_{cr} [36–39]. The theoretical predictions were afterward confirmed experimentally by means of autocorrelation measurements [40]. Pulse splitting was further investigated by recording cross-correlation functions which confirmed the numerically predicted asymmetry between the intensities of the split sub-pulses [41]. The detailed amplitude structure and phase information of the split sub-pulses was retrieved from the frequency-resolved optical gating (FROG) technique [42], eventually establishing the general link between pulse splitting and SC generation [43].

On the basis of these findings and numerical simulation results, a temporal scenario of SC generation in normally dispersive media was proposed [44], highlighting the role of pulse splitting events. Self-phase modulation broadens the pulse spectrum and produces a nonlinear frequency modulation (chirp) in which red-shifted and blue-shifted frequencies are generated at the leading and trailing parts of the pulse, respectively. Pulse splitting at the nonlinear focus produces two sub-pulses with shifted carrier frequencies. The split sub-pulses move in opposite directions in the local frame of the input pulse, as illustrated in Fig. 2.6a. The velocity difference between a pulse peak and its tails, due to the refractive index dependence on the intensity, induces sharp intensity gradients (optical shocks) in the temporal profiles of the sub-pulses. Pulse splitting is thus immediately followed by an explosive broadening of the spectrum (Fig. 2.6a), produced by the latter self-steepening effect.

The asymmetry in experimentally measured shapes of the SC spectra can be qualitatively explained by the rather different self-steepening effects experienced by the split sub-pulses. In the near-infrared spectral range, under typical focusing conditions, a particularly steep edge is formed at the trailing edge of the trailing sub-pulse, giving rise to a broad blue-shifted pedestal in the SC spectrum. In contrast, the self-steepening of the leading front of the leading sub-pulse is less significant, resulting in a rather modest red-shifted spectral broadening. Dedicated measurements of the spectral content of the split sub-pulses provided an experimental verification of the connections between the leading and trailing sub-pulses and the red-shifted and the blue-shifted spectral broadening [45]. The universality of this scenario of SC generation in normally dispersive media, accompanied by pulse splitting, is confirmed by the apparent similarity of the spectral shapes of SC generated in various nonlinear media.

3.4 Supercontinuum Generation in the Region of Anomalous GVD

In 1990, Silberberg extended the concept of self-trapping of light beams to optical pulses propagating in dielectric media in the region of anomalous GVD [46]. By means of a space-time coordinate transformation reducing to a one dimensional problem with spherical symmetry the search for propagation invariant solutions to the nonlinear Schrödinger equation, Silberberg found a three-dimensional (spatiotemporal) localized wavepacket that does not spread in space or time, which constitutes the counterpart of the purely spatial Townes mode and is called a *light bullet* [46]. From this discovery, a qualitatively different temporal scenario of self-focusing and femtosecond filamentation was foreseen in the range of anomalous GVD, suggesting that the interplay between self-focusing, self-phase modulation, and anomalous GVD may lead to simultaneous shrinking of the input wave packet in spatial and temporal dimensions, potentially giving rise to self-compressed three-dimensional self-trapped pulses [47]. Here, new red-shifted and blue-shifted frequencies, that are generated by the self-phase modulation on the ascending (leading) and descending (trailing) edges of the pulse, respectively, are swept back to the peak of the pulse, instead of being dispersed as in the case of normal GVD.

A truly three-dimensional self-trapped pulse undergoing filamentation requires that beam collapse due to Kerr self-focusing and pulse compression due to group velocity dispersion follow the same rate in order to reach minimal dimensions at the nonlinear focus. Since material parameters are not controllable, this is hardly possible except in specific conditions for the pulse peak power, duration and beam width. Nevertheless Moll et al. [47] identified a more generic scenario initiated by a two-dimensional self-focusing stage and collapse, arrested by the formation of a plasma, and followed by a continuous transfer of energy into the collapse region due to the anomalous GVD. This resulted in a remarkable (almost 10 times) increase of the filamentation length [47] before the beam eventually defocuses. The feasibility of self-compressed objects was confirmed by numerical simulations of the earliest [47, 48] and more recent experiments [49, 50], which predicted that pulse self-compression down to a single optical cycle is possible.

The development of high-peak-power near- and mid-infrared ultrashort-pulse laser sources, which are exclusively based on the optical parametric amplification gave an experimental access to study filamentation phenomena in the range of anomalous GVD of wide bandgap dielectrics and even semiconductors, whose zero GVD wavelengths are located deeply in the mid-infrared. An ultrabroadband SC emission [51] was observed by launching femtosecond pulses at 1.55 μm in fused silica. A more recent study demonstrated filamentation of incident pulses with much longer wavelength (3.1 μm) in a YAG crystal, yielding more than three octave-spanning SC spectrum with unprecedented wavelength coverage from the ultraviolet to the mid-infrared [52]. Eventually, simultaneous time and space compression was demonstrated to favor a new type of filamentation, which produces propagation invariant three-dimensional self-compressed light bullets that preserve a narrow beam diameter and a short pulsewidth over a considerable distance in a nonlinear dispersive medium [50]. The formation of self-compressed spatiotemporal light bullets was experimentally observed in various nonlinear media, such as fused silica, sapphire and BBO, and under a variety of operating conditions [53–57]. Light bullets in the form of localized wavepackets as originally proposed by Silberberg are subject to the same instabilities as the Townes mode [58]. For this reason, it came as no surprise that experimentally observed light bullets were found to posses all the attributes of weakly localized conical waves [54], thereby providing a universal description of *self-trapped filaments* in normal and anomalous dispersion regions.

Figure 2.6c shows a numerical example illustrating the formation and propagation dynamics of the self-compressed spatiotemporal light bullet in a sapphire crystal, which is accompanied by the generation of an ultrabroadband SC (Fig. 2.6f), which emerges at the point where pulse self-compression occurs.

3.5 Supercontinuum Generation Under Zero GVD

For commonly used wide bandgap dielectric materials, wavelengths corresponding to zero GVD lie in the near-infrared spectral range between 1 and 2 μm, see Table 4.1 in Chap. 4. To some extent, the combined properties of both normal and anomalous GVD characterize the near-zero GVD regime. In the time domain, the input pulse still undergoes splitting at the nonlinear focus, as illustrated in Fig. 2.6b, and the post-collapse dynamics are essentially similar to those observed in the case of normal GVD shown in Fig. 2.6a. This is due to the fact that not only GVD can induce pulse splitting but also nonlinear effects such as multiphoton absorption by acting on the most intense part of the pulse [59]. Close to zero GVD, nonlinear absorption indeed becomes the main responsible for pulse splitting and as it acts in the same way on the initially symmetric leading and trailing edges of the propagating pulse, the spectrum is broadened much more symmetrically than in the two previous propagation regimes. Experimental measurements show that pulse splitting prevails even in the case of weak anomalous GVD [60], where the amount of material dispersion is too small to compress the spectrally broadened pulse.

Fig. 3.4 Numerically
simulated axial
supercontinuum spectra after
4 mm-thick sapphire crystal
in the cases of **a** normal
GVD, **b** zero GVD, **c**
anomalous GVD. The input
spectra are shown by the
dashed curves

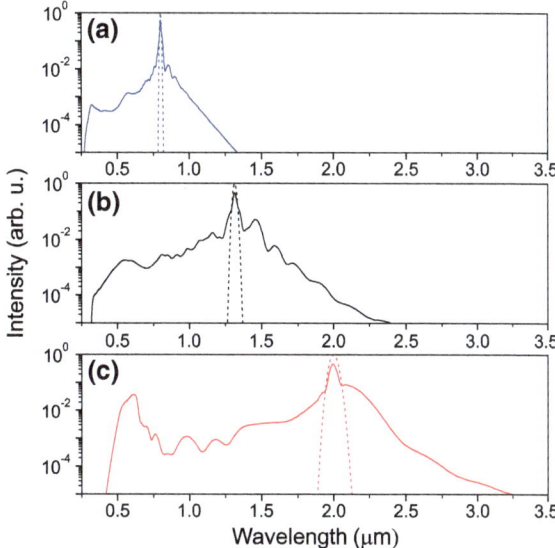

Figure 3.4 presents a comparison of the numerically simulated axial SC spectra generated by self-focusing and filamentation of 100 fs pulses in 4 mm-thick sapphire crystal, in the regimes of normal (the input wavelength 800 nm), zero (1.3 μm) and anomalous (2.0 μm) GVD.

3.6 Conical Emission

As already underlined, observations of filamentation in various experimental conditions indicated that the propagation dynamics excites nonlinear conical waves with universal features, which are propagation invariant wavepackets likely to play the role of attractors for the filamentation dynamics. The above filamentation and SC generation scenarios could be generalized by employing the effective three-wave mixing model, which provides the unified picture, connecting the spectral broadening on the propagation axis with colored conical emission [61, 62], which is perhaps the most striking and visually perceptible evidence of SC generation in bulk media. The spectral content and the angular distribution of the scattered waves satisfy phase-matching conditions, which are defined by the chromatic dispersion, thereby providing a particular dispersion-defined angular landscape of scattered frequencies [63]. Experimentally, these landscapes could be retrieved by measuring the SC spectrum with an imaging spectrometer [64]. Figure 3.5a–c show the experimentally measured angularly resolved SC spectra in water, which exhibit qualitatively different patterns of conical emission in the range of normal and anomalous GVD [65]. More specifically, in the range of normal GVD, off-axis (conical) tails emerge on both the blue

Fig. 3.5 Experimentally measured angle-resolved spectra around the incident wavelengths of 527 nm and 1.055 μm that fall into the ranges of **a** normal GVD, **b** anomalous GVD of water, respectively. Adapted from [65]. **c** The entire angle-resolved SC spectrum in water as excited with 1.055 μm input pulses. Adapted from [66]. **d** The angle-resolved SC spectrum as excited with 800 nm input pulses in sapphire. The white solid curves indicate the best fits obtained using the X-wave relation. Adapted from [67]. **a** and **b** reprinted by permission of the Optical Society of America, **c** and **d** reprinted by permission of American Physical Society

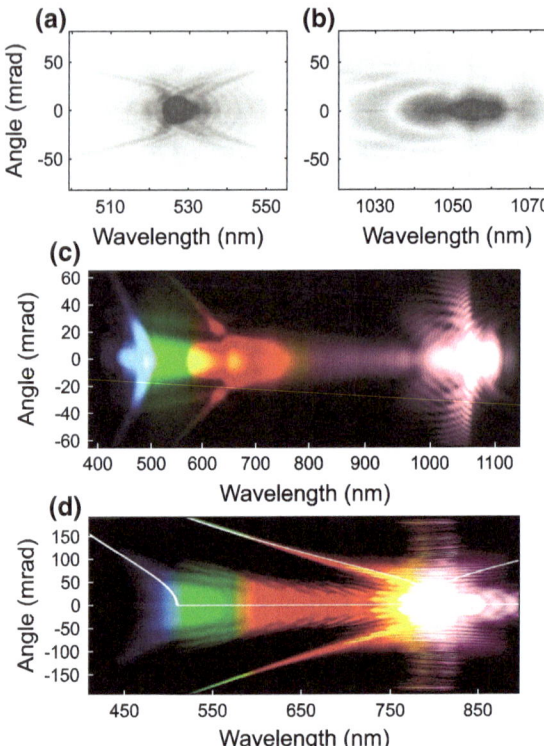

and red-shifted sides of the input wavelength, forming a distinct X-shaped pattern of conical emission, as shown in Fig. 3.5a. In contrast, in the range of anomalous GVD, conical emission pattern develops a multiple annular, or O-shaped structure around the input wavelength, as illustrated in Fig. 3.5b. The entire angle-resolved SC spectrum produced by filamentation of 1055 nm laser pulses, whose wavelengths fall into the range of anomalous GVD of water, is presented in Fig. 3.5c. It consists of multiple annular structures around the carrier wavelength and a distinct V-shaped tail in the visible spectral range [66]. Figure 3.5d illustrates the entire angle-resolved SC spectrum in sapphire, as generated by 800 nm pulses, in the range of normal GVD [67].

The shapes of angle-resolved SC spectra are universal for any other nonlinear medium. These findings led to the interpretation of femtosecond filaments as conical waves, assuming that the input wave packet will try to evolve toward a final stationary state that has the form of either an X-wave in the range of normal GVD or an O-wave in the range of anomalous GVD. Nonlinear X-waves [59, 68] and O-waves [21] are named because of their evident X-like and O-like shapes, respectively, which appear in both the near and the far fields. Moreover, the interpretation of light filaments in the framework of conical waves readily explains the distinctive propagation features of light filaments such as sub-diffractive propagation in free space [54, 67] and self-reconstruction after hitting physical obstacles [12, 13, 69], which are universal and regardless of the sign of material GVD, and which were verified experimentally as well. Therefore, all subsequent features of the filament propagation in the regime of normal GVD, i.e., pulse splitting, conical emission, and any nonlinear interactions, may be interpreted assuming the pulses as spontaneously occurring nonlinear X-waves [70]. Consequently, the formation and propagation features of spatiotemporal light bullets in the regime of anomalous GVD, may be interpreted in terms of nonlinear O-waves [54].

References

1. Braun, A., Korn, G., Liu, X., Du, D., Squier, J., Mourou, G.: Self-channeling of high-peak-power femtosecond laser pulses in air. Opt. Lett. **20**, 73–75 (1995)
2. Tzortzakis, S., Sudrie, L., Franco, M., Prade, B., Mysyrowicz, A., Couairon, A., Bergé, L.: Self-guided propagation of ultrashort IR laser pulses in fused silica. Phys. Rev. Lett. **87**, 213902 (2001)
3. Hercher, M.: Laser-induced change in transparent media. J. Opt. Soc. Am. **54**, 563 (1964)
4. Couairon, A., Mysyrowicz, A.: Femtosecond filamentation in transparent media. Phys. Rep. **441**, 47–190 (2007)
5. Nibbering, E.T.J., Curley, P.F., Grillon, G., Prade, B.S., Franco, M.A., Salin, F., Mysyrowicz, A.: Conical emission from self-guided femtosecond pulses in air. Opt. Lett. **21**, 62–64 (1996)
6. Yang, J., Mu, G.: Multi-dimensional observation of white-light filaments generated by femtosecond laser pulses in condensed medium. Opt. Express **15**, 4943–4952 (2007)
7. Tzortzakis, S., Franco, M.A., André, Y.-B., Chiron, A., Lamouroux, B., Prade, B.S., Mysyrowicz, A.: Formation of a conducting channel in air by self-guided femtosecond laser pulses. Phys. Rev. E **60**, R3505 (1999)
8. Minardi, S., Gopal, A., Tatarakis, M., Couairon, A., Tamošauskas, G., Piskarskas, R., Dubietis, A., Di Trapani, P.: Time-resolved refractive index and absorption mapping of light-plasma filaments in water. Opt. Lett. **33**, 86–88 (2008)
9. Papazoglou, D.G., Tzortzakis, S.: In-line holography for the characterization of ultrafast laser filamentation in transparent media. Appl. Phys. Lett. **93**, 041120 (2008)
10. Papazoglou, D.G., Tzortzakis, S.: Physical mechanisms of fused silica restructuring and densification after femtosecond laser excitation. Opt. Mater. Express **1**, 625–632 (2011)
11. Courvoisier, F., Boutou, V., Kasparian, J., Salmon, E., Méjean, G., Yu, J., Wolf, J.-P.: Ultraintense light filaments transmitted through clouds. Appl. Phys. Lett. **83**, 213–215 (2003)
12. Dubietis, A., Gaižauskas, E., Tamošauskas, G., Di Trapani, P.: Light filaments without self-channeling. Phys. Rev. Lett. **92**, 253903 (2004)
13. Dubietis, A., Kučinskas, E., Tamošauskas, G., Gaižauskas, E., Porras, M.A., Di Trapani, P.: Self-reconstruction of light filaments. Opt. Lett. **29**, 2893–2895 (2004)
14. Kolesik, M., Moloney, J.V.: Self-healing femtosecond light filaments. Opt. Lett. **29**, 590–592 (2004)
15. Liu, W., Gravel, J.-F., Théberge, F., Becker, A., Chin, S.L.: Background reservoir: its crucial role for long-distance propagation of femtosecond laser pulses in air. Appl. Phys. B **80**, 857–860 (2005)
16. Liu, W., Thèberge, F., Arévalo, E., Gravel, J.-F., Becker, A., Chin, S.L.: Experiments and simulations on the energy reservoir effect in femtosecond light filaments. Opt. Lett. **30**, 2602–2604 (2005)
17. Chiao, R.Y., Garmire, E., Townes, C.H.: Self-trapping of optical beams. Phys. Rev. Lett. **13**, 479–482 (1964)

18. Saari, P., Reivelt, K.: Evidence of X-shaped propagation-invariant localized light waves. Phys. Rev. Lett. **79**, 4135–4138 (1997)

19. Lu, J.Y., Greenleaf, J.F.: Nondiffracting x waves-exact solutions for free-space scalar wave equation and their finite aperture realizations. IEEE Trans. Ultrasonics, Ferroelectr. Freq. Control **39**, 19–31 (1992)

20. Porras, M.A., Parola, A., Faccio, D., Dubietis, A., Di Trapani, P.: Nonlinear unbalanced Bessel Beams: Stationary conical waves supported by nonlinear losses. Phys. Rev. Lett. **93**, 153902 (2004)

21. Porras, M.A., Parola, A., Di Trapani, P.: Nonlinear unbalanced O waves: nonsolitary, conical light bullets in nonlinear dissipative media. J. Opt. Soc. Am. B **22**, 1406–1413 (2005)

22. Conti, C., Trillo, S., Di Trapani, P., Valiulis, G., Piskarskas, A., Jedrkiewicz, O., Trull, J.: Nonlinear electromagnetic X waves. Phys. Rev. Lett. **90**, 170406 (2003)

23. Polesana, P., Dubietis, A., Porras, M.A., Kučinskas, E., Faccio, D., Couairon, A., Di Trapani, P.: Near-field dynamics of ultrashort pulsed Bessel beams in media with Kerr nonlinearity. Phys. Rev. E **73**, 056612 (2006)

24. Valiulis, G., Kilius, J., Jedrkiewicz, O., Bramati, A., Minardi, S., Conti, C., Trillo, S., Piskarskas, A., Di Trapani, P.: Space-time nonlinear compression and three-dimensional complex trapping in normal dispersion. In: OSA Trends in Optics and Photonics (TOPS), Technical Digest of the Quantum Electronics and Laser Science Conference (QELS 2001), vol. 57. Optical Society of America, Washington DC, pp. QPD1012 (2001)

25. Di Trapani, P., Valiulis, G., Piskarskas, A., Jedrkiewicz, O., Trull, J., Conti, C., Trillo, S.: Spontaneously generated X-shaped light bullets. Phys. Rev. Lett. **91**, 093904 (2003)

26. Faccio, D., Couairon, A., Di Trapani, P.: Conical Waves, Filaments and Nonlinear Filamentation Optics Aracne Rome (2007)

27. Kandidov, V.P., Kosareva, O.G., Shlenov, S.A.: Influence of transient self-defocusing on the propagation of high-power femtosecond laser pulses in gases under ionisation conditions. Quant. Electron. **21**, 971–977 (1994)

28. Mlejnek, M., Wright, E.M., Moloney, J.V.: Dynamic spatial replenishment of femtosecond pulses propagating in air. Opt. Lett. **23**, 382–384 (1998)

29. Chiron, A., Lamouroux, B., Lange, R., Ripoche, J.-F., Franco, M., Prade, B., Bonnaud, G., Riazuelo, G., Mysyrowicz, A.: Numerical simulations of the nonlinear propagation of femtosecond optical pulses in gases. Eur. Phys. J. D **6**, 383–396 (1999)

30. Kolesik, M., Moloney, J.V.: Nonlinear optical pulse propagation simulation: From Maxwell's to unidirectional equations. Phys. Rev. E **70**, 036604 (2004)

31. Couairon, A., Brambilla, E., Corti, T., Majus, D., de O., Ramírez-Góngora, J., Kolesik, M.: Practitioner's guide to laser pulse propagation models and simulation. Eur. Phys. J. Special Topics **199**, 5–76 (2011)

32. Brabec, T., Krausz, F.: Nonlinear optical pulse propagation in the single-cycle regime. Phys. Rev. Lett. **78**, 3282–3285 (1997)

33. Couairon, A., Kosareva, O.G., Panov, N.A., Shipilo, D.E., Andreeva, V.A., Jukna, V., Nesa, F.: Propagation equation for tight-focusing by a parabolic mirror. Opt. Express **23**, 31240–31252 (2015)

34. Husakou, A.V., Herrmann, J.: Supercontinuum generation of higher-order solitons by fission in photonic crystal fibers. Phys. Rev. Lett. **87**, 203901 (2001)

35. Kolesik, M., Moloney, J.V.: Modeling and simulation techniques in extreme nonlinear optics of gaseous and condensed media. Rep. Prog. Phys. **77**, 016401 (2014)

36. Chernev, P., Petrov, V.: Self-focusing of light pulses in the presence of normal group-velocity dispersion. Opt. Lett. **17**, 172–174 (1992)

37. Rothenberg, J.E.: Pulse splitting during self-focusing in normally dispersive media. Opt. Lett. **17**, 583–585 (1992)

38. Rothenberg, J.E.: Space-time focusing: breakdown of the slowly varying envelope approximation in the self-focusing of femtosecond pulses. Opt. Lett. **17**, 1340–1342 (1992)

39. Fibich, G., Papanicolaou, G.C.: Self-focusing in the presence of small time dispersion and nonparaxiality. Opt. Lett. **22**, 1397–1399 (1997)

40. Ranka, J.K., Schirmer, R.W., Gaeta, A.L.: Observation of pulse splitting in nonlinear dispersive media. Phys. Rev. Lett. **77**, 3783–3786 (1996)
41. Ranka, J.K., Gaeta, A.L.: Breakdown of the slowly varying envelope approximation in the self-focusing of ultrashort pulses. Opt. Lett. **23**, 534–536 (1998)
42. Diddams, S.A., Eaton, H.K., Zozulya, A.A., Clement, T.S.: Amplitude and phase measurements of femtosecond pulse splitting in nonlinear dispersive media. Opt. Lett. **23**, 379–381 (1998)
43. Zozulya, A.A., Diddams, S.A., Van Engen, A.G., Clement, T.S.: Propagation dynamics of intense femtosecond pulses: multiple splittings, coalescence, and continuum generation. Phys. Rev. Lett. **82**, 1430–1433 (1999)
44. Gaeta, A.L.: Catastrophic collapse of ultrashort pulses. Phys. Rev. Lett. **84**, 3582–3585 (2000)
45. Gaeta, A.L.: Spatial and temporal dynamics of collapsing ultrashort laser pulses. Top. Appl. Phys. **114**, 399–412 (2009)
46. Silberberg, Y.: Collapse of optical pulses. Opt. Lett. **15**, 1282–1284 (1990)
47. Moll, K.D., Gaeta, A.L.: Role of dispersion in multiple-collapse dynamics. Opt. Lett. **29**, 995–997 (2004)
48. Liu, J., Li, R., Xu, Z.: Few-cycle spatiotemporal soliton wave excited by filamentation of a femtosecond laser pulse in materials with anomalous dispersion. Phys. Rev. A **74**, 043801 (2006)
49. Chekalin, S.V., Kompanets, V.O., Smetanina, E.O., Kandidov, V.P.: Light bullets and super-continuum spectrum during femtosecond pulse filamentation under conditions of anomalous group-velocity dispersion in fused silica. Quantum Electron. **43**, 326–331 (2013)
50. Durand, M., Jarnac, A., Houard, A., Liu, Y., Grabielle, S., Forget, N., Durécu, A., Couairon, A., Mysyrowicz, A.: Self-guided propagation of ultrashort laser pulses in the anomalous dispersion region of transparent solids: a new regime of filamentation. Phys. Rev. Lett. **110**, 115003 (2013)
51. Saliminia, A., Chin, S.L., Vallée, R.: Ultra-broad and coherent white light generation in silica glass by focused femtosecond pulses at 1.5 μm. Opt. Express **13**, 5731–5738 (2005)
52. Silva, F., Austin, D.R., Thai, A., Baudisch, M., Hemmer, M., Faccio, D., Couairon, A., Biegert, J.: Multi-octave supercontinuum generation from mid-infrared filamentation in a bulk crystal. Nature Commun. **3**, 807 (2012)
53. Smetanina, E.O., Kompanets, V.O., Dormidonov, A.E., Chekalin, S.V., Kandidov, V.P.: Light bullets from near-IR filament in fused silica. Laser Phys. Lett. **10**, 105401 (2013)
54. Majus, D., Tamošauskas, G., Gražulevičiūtė, I., Garejev, N., Lotti, A., Couairon, A., Faccio, D., Dubietis, A.: Nature of spatiotemporal light bullets in bulk Kerr media. Phys. Rev. Lett. **112**, 193901 (2014)
55. Gražulevičiūtė, I., Šuminas, R., Tamošauskas, G., Couairon, A., Dubietis, A.: Carrier-envelope phase-stable spatiotemporal light bullets. Opt. Lett. **40**, 3719–3722 (2015)
56. Chekalin, S.V., Dokukina, A.E., Dormidonov, A.E., Kompanets, V.O., Smetanina, E.O., Kandidov, V.P.: Light bullets from a femtosecond filament. J. Phys. B **48**, 094008 (2015)
57. Šuminas, R., Tamošauskas, G., Valiulis, G., Dubietis, A.: Spatiotemporal light bullets and supercontinuum generation in β-BBO crystal with competing quadratic and cubic nonlinearities. Opt. Lett. **41**, 2097–2100 (2016)
58. Porras, M.A., Parola, A., Faccio, D., Couairon, A., Di Trapani, P.: Light-filament dynamics and the spatiotemporal instability of the Townes profile. Phys. Rev. A **76**, 011803(R) (2007)
59. Couairon, A., Gaižauskas, E., Faccio, D., Dubietis, A., Di Trapani, P.: Nonlinear X-wave formation by femtosecond filamentation in Kerr media. Phys. Rev. E **73**, 016608 (2006)
60. Gražulevičiūtė, I., Garejev, N., Majus, D., Jukna, V., Tamošauskas, G., Dubietis, A.: Filamentation and light bullet formation dynamics in solid-state dielectric media with weak, moderate and strong anomalous group velocity dispersion. J. Opt. **18**, 025502 (2016)
61. Kolesik, M., Katona, G., Moloney, J.V., Wright, E.M.: Physical factors limiting the spectral extent and band gap dependence of supercontinuum generation. Phys. Rev. Lett. **91**, 043905 (2003)
62. Kolesik, M., Katona, G., Moloney, J.V., Wright, E.M.: Theory and simulation of supercontinuum generation in transparent bulk media. Appl. Phys. B **77**, 185–195 (2003)

63. Kolesik, M., Wright, E.M., Moloney, J.V.: Interpretation of the spectrally resolved far field of femtosecond pulses propagating in bulk nonlinear dispersive media. Opt. Express **13**, 10729–10741 (2005)

64. Faccio, D., Di Trapani, P., Minardi, S., Bramati, A., Bragheri, F., Liberale, C., Degiorgio, V., Dubietis, A., Matijosius, A.: Far-field spectral characterization of conical emission and filamentation in Kerr media. J. Opt. Soc. Am. B **22**, 862–869 (2005)

65. Porras, M.A., Dubietis, A., Kučinskas, E., Bragheri, F., Degiorgio, V., Couairon, A., Faccio, D., Di Trapani, P.: From X- to O-shaped spatiotemporal spectra of light filaments in water. Opt. Lett. **30**, 3398–3400 (2005)

66. Faccio, D., Averchi, A., Lotti, A., Kolesik, M., Moloney, J.V., Couairon, A., Di Trapani, P.: Generation and control of extreme blueshifted continuum peaks in optical Kerr media. Phys. Rev. A **78**, 033825 (2008)

67. Faccio, D., Clerici, M., Averchi, A., Lotti, A., Jedrkiewicz, O., Dubietis, A., Tamošauskas, G., Couairon, A., Bragheri, F., Papazoglou, D., Tzortzakis, S., Di Trapani, P.: Few-cycle laser-pulse collapse in Kerr media: The role of group-velocity dispersion and X-wave formation. Phys. Rev. A **78**, 033826 (2008)

68. Kolesik, M., Wright, E.M., Moloney, J.V.: Dynamic nonlinear X waves for femtosecond pulse propagation in water. Phys. Rev. Lett. **92**, 253901 (2004)

69. Gražulevičiūtė, I., Tamošauskas, G., Jukna, V., Couairon, A., Faccio, D., Dubietis, A.: Self-reconstructing spatiotemporal light bullets. Opt. Express **22**, 30613–30622 (2014)

70. Faccio, D., Porras, M.A., Dubietis, A., Bragheri, F., Couairon, A., Di Trapani, P.: Conical emission, pulse splitting, and X-wave parametric amplification in nonlinear dynamics of ultrashort light pulses. Phys. Rev. Lett. **96**, 193901 (2006)

Part II
Overview of the Experimental Results

Chapter 4
General Practical Considerations

Supercontinuum generation in transparent bulk media results from femtosecond filamentation, which involves a complex interplay among linear (diffraction and GVD) and nonlinear effects (self-focusing, self-phase modulation, pulse-front steepening, generation of optical shocks, multiphoton absorption/ionization, free-electron plasma generation, etc.), which become coupled in space and time. Despite complex fundamental issues, the practical setup for supercontinuum generation is amazingly simple. It involves just a focusing lens, a piece of suitable nonlinear material and a collimating lens. However, to generate stable and reproducible SC, a number of important practical issues, which are universal and hold for any nonlinear medium and at any input wavelength in the optical range should be taken into consideration. This chapter provides an overview of relevant parameters of most frequently used nonlinear materials, discusses the issues related to beam focusing geometry, and the numerical aperture in particular, origins of SC instabilities, the effects on SC generation related to filament refocusing, and the pros and cons of SC generation in the single and multiple filamentation regimes.

4.1 Materials

Proper choice of the nonlinear medium is the primary issue that guarantees the generation of stable and reproducible SC spectrum within a desired wavelength range. In general, the spectral extent of the SC is defined essentially by the laser wavelength and by linear and nonlinear properties of the medium, such as energy bandgap, transparency range, nonlinear index of refraction, and the sign and amount of chromatic dispersion, which possess fundamental mutual relationships. The experimentally established dependence of the SC spectral width on the material bandgap suggests that the broadest SC spectra with the largest blueshifts from the carrier wavelength

A. Dubietis and A. Couairon, *Ultrafast Supercontinuum Generation in Transparent Solid-State Media*, SpringerBriefs in Physics, https://doi.org/10.1007/978-3-030-14995-6_4

could be attained in wide bandgap dielectrics [1, 2]. More precisely, a universal parameter, which relates the material and laser parameters, i.e., the material bandgap and the incident photon energy, is expressed through the order of multiphoton absorption: $K = \langle U_g/\hbar\omega_0 \rangle + 1$, where U_g is the bandgap and $\hbar\omega_0$ is the photon energy that is inversely proportional to the laser wavelength. A large collection of experimental data shows that the threshold value for SC generation is $K \geq 3$, so defining the suitability of the particular nonlinear material for SC generation with a given pump wavelength. This implies that no SC generation, or at least no appreciable spectral broadening could be achieved in the regime of two-photon absorption.

Numerical studies demonstrated that besides the order of multiphoton absorption, chromatic dispersion of the material is an equally important player, which determines attainable both blue and redshifts of the SC spectrum. First, it was shown that larger spectral broadening on the short-wavelength side is produced in the materials with lower chromatic dispersion [3]. Second, a more recent numerical study of SC generation in various nonlinear materials using mid-infrared laser pulses uncovered that the zero GVD wavelength may serve as a reasonably good indicator for attainable redshift of the SC spectrum [4]. These findings suggest that SC spectra with remarkable redshifts could be generated in the nonlinear materials whose zero GVD wavelengths are located in the mid-infrared.

Table 4.1 presents relevant linear and nonlinear parameters of widely used nonlinear dielectric materials, water and some popular nonlinear crystals with second-order nonlinearity (BBO and KDP) that are used for SC generation. Modern amplified laser sources, such as Ti:sapphire and Yb-doped lasers (Yb:KGW, Yb:KYW, Yb:fiber,

Table 4.1 Linear and nonlinear parameters of basic dielectric media used for supercontinuum generation. U_g is the energy bandgap, the transmission range is defined at 10% transmission level in a 1 mm thick sample, n_0 and n_2 are linear and nonlinear refractive indexes, respectively, and are given for $\lambda = 800$ nm, λ_0 is the zero GVD wavelength. Reproduced from [5] with permission

Material	U_g eV	Transmittance µm	n_2 $\times 10^{-16}$ cm^2/W	n_0	λ_0 µm
LiF	13.6	0.12–6.6	0.81	1.39	1.23
CaF$_2$	10	0.12–10	1.3	1.43	1.55
Al$_2$O$_3$	9.9	0.19–5.2	3.1	1.76	1.31
BaF$_2$	9.1	0.14–13	1.91	1.47	1.93
SiO$_2$ (FS)	9.0	0.18–3.5	2.4	1.45	1.27
KDP	7.0	0.18–1.55	2.0	1.50	0.98
H$_2$O	6.9	0.18–1.3	5.7	1.33	1.0
YAG	6.5	0.21–5.2	6.2	1.82	1.60
β-BBO	6.2	0.19–3.5	5.2	1.66	1.49
BK7	4.28	0.3–2.5	3.75	1.51	1.32
KGW	4.05	0.3–5	11	2.02	2.2
YVO$_4$	3.8	0.35–4.8	15	1.97	

etc.) provide high energy ultrashort pulses with durations ranging from a few tens to a few hundreds of femtoseconds, with carrier wavelengths around 800 nm and 1.03 μm, respectively, which fall into the range of normal GVD of the vast majority of listed materials. Second and third harmonic generation by these laser sources afford obtaining femtosecond pulses in the visible and ultraviolet spectral range. Frequency down-conversion techniques based on the optical parametric amplification and difference frequency generation, provide a great flexibility in the choice of accessible wavelengths in the near- and mid-infrared, where the GVD of dielectric materials is anomalous. Moreover, the availability of ultrashort pulses with longer pump wavelengths greatly extends the choice of suitable nonlinear materials for SC generation beyond those listed in Table 4.1, and include narrow bandgap dielectrics, soft glasses and semiconductors, which posses remarkably broad transmission windows in the mid-infrared, as will be discussed in more detail in Chap. 5.

Experimentally, a light filament and consequently, a SC is produced when the incident beam power exceeds the critical power for self-focusing P_{cr} just by a relatively small fraction (several tens of percents). The particular value of the input beam power slightly depends on the external focusing geometry and the length of the nonlinear material. In the near-infrared spectral range, the typical values of P_{cr} in wide bandgap dielectric media are of the order of several MW, that are easily achieved with femtosecond laser pulses having energies in the range from a few microjoules to a few hundreds of nanojoules. Figure 4.1 illustrates the calculated critical power for self-focusing in various dielectric materials for the input wavelengths of 800 nm and 1030 nm, which are emitted by commonly used Ti:sapphire and Yb-doped femtosecond lasers, respectively. Notice that P_{cr} scales as λ^2, so filamentation and SC generation threshold for ultraviolet pulses is notably reduced, while the opposite is true for pulses in the mid-infrared.

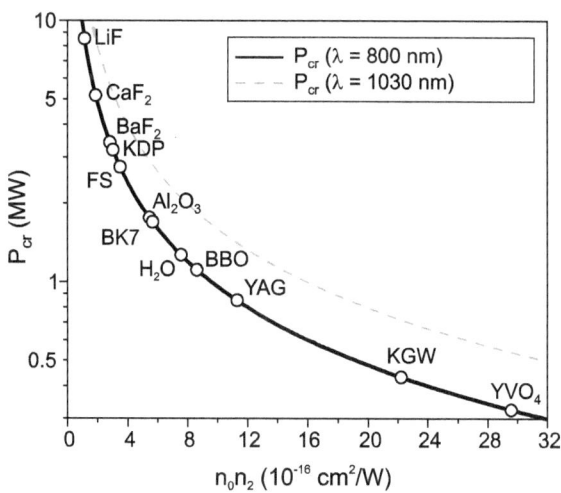

Fig. 4.1 Critical power for self-focusing in various dielectric materials calculated for Ti:sapphire (800 nm, solid curve) and Yb-doped (1030 nm, dashed curve) laser wavelengths. Reproduced from [5] with permission

4.2 External Focusing

External focusing of the input beam plays an important role concerning various practical aspects of SC generation. Proper choice of the external focusing geometry allows the setting of safe operating conditions and avoiding unwanted effects such as optical damage of the material and enables to optimize the SC generation setup for achieving stable and reproducible operation.

The optical damage of the medium is the major limiting factor to SC generation, which dictates the focusing geometry of the incident pump beam. Typically, the optical damage occurs either within the bulk at the vicinity of the nonlinear focus or at the output face of the nonlinear medium, where the beam diameter shrinks considerably due to self-focusing, and the beam is transformed into a narrow and intense filament. Table 4.2 provides the experimental single-shot surface damage threshold values for some basic dielectric materials, as measured with 500 fs pulses at 1030 nm [6]. Note that, the thresholds for optical damage are provided for a single-shot irradiation, and do not account for the optical degradation of the materials (due to color center formation, etc.) exposed to multiple repetitive pulses. In the femtosecond range, the optical damage threshold scales with the material bandgap and input pulse duration t_p, which for oxides is expressed by the empirical relation [7]:

$$F_{th}(U_g, t_p) = (c_1 + c_2 U_g)t_p^\kappa, \tag{4.1}$$

where $c_1 = -0.16 \text{ J cm}^{-2} \text{ fs}^{-\kappa}$, $c_2 = 0.074 \text{ J cm}^{-2} \text{ fs}^{-\kappa} \text{ eV}^{-1}$, and $\kappa = 0.3$.

Experimental and numerical studies of the laser-induced damage threshold of various optical materials have been performed in a range of wavelengths from the near-infrared to the ultraviolet, showing rather complex dependences on the material bandgap and incident wavelength [8]. As a general rule, all the materials are prone to optical damage in the ultraviolet, and the damage threshold increases with increasing incident wavelength.

Considering the above, setting proper external focusing condition (the numerical aperture of the input beam, NA) is of major importance as it will prevent optical damage of the material in the SC generation process [9, 10]. As a general example, Fig. 4.2 shows the experimentally measured threshold energies for SC generation and optical damage in fused silica as functions of the numerical aperture [9]. In the high NA regime (NA > 0.25), the optical damage occurs for the input pulse energies below the energy corresponding to the critical power of self-focusing, so no SC generation under such focusing condition is observed. Most of the pulse energy is thus

Table 4.2 Single-shot optical damage thresholds for some dielectric materials, measured with 500 fs pulses at 1030 nm [6]

Material	YAG	Al_2O_3	SiO_2	CaF_2	LiF	BaF_2
F_{th}, J/cm^2	7.5	5.36	4.85	5.17	4.43	2.91

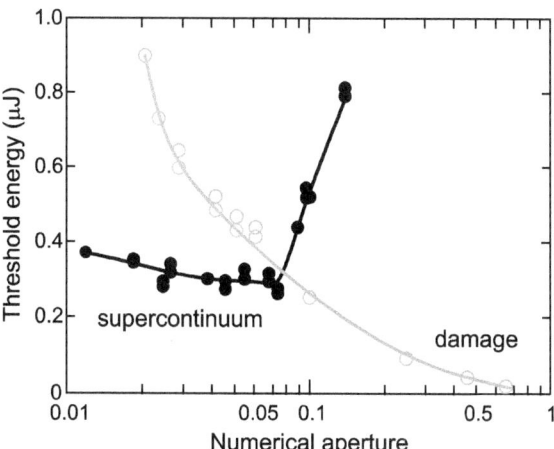

Fig. 4.2 Energy thresholds for the optical damage (open circles) and supercontinuum generation (filled circles) measured in UV-grade fused silica with 60 fs, 800 nm laser pulses versus numerical aperture. The curves serve as guides for the eye. Adapted from [9] and reprinted by permission from the Optical Society of America

deposited into the material at the focal volume through the multiphoton absorption and subsequent absorption by free-electron plasma via the inverse Bremsstrahlung effect, representing a laser–matter interaction regime that is commonly exploited for micromachining of transparent bulk materials. In the NA range from 0.15 to 0.05, the threshold energies for SC generation and optical damage are very close. In particular, for NA < 0.1, the SC is generated without the optical damage in a single-shot regime, however, the damage accumulates under multiple shot exposure, causing the SC to disappear over time. Finally, with NA below 0.05, it is still possible to damage the medium, but only by increasing the input pulse energies significantly above the threshold for SC generation, hence constituting the "safe" operating condition for SC generation.

Typically, the incident pump beam is externally focused to a diameter of 30–100 μm, which guarantees the location of the nonlinear focus inside the nonlinear medium of several mm thickness, as could be fairly correctly estimated from Eq. (2.7). In practice, the position of the nonlinear focus could be rather precisely estimated by monitoring the filament-induced luminescence trace from the side view of the nonlinear material. Filament-induced luminescence originates from the relaxation of electron excitations by the multiphoton absorption and inverse Bremsstrahlung effect. The perceptible color of filament-induced luminescence depends on the material, but is almost independent on the pump wavelength. For instance, YAG and sapphire produce broadband ultraviolet luminescence, associated with exciton and antisite defect emissions in YAG and F^+ center emissions in sapphire, whose long-wavelength side extends into the visible range, appearing as a faint blue-violet trace seen by the naked eye, see e.g., [11, 12]. Impurities or dopants (e.g., Ti ions in sapphire, Nd, Yb, etc., ions in YAG) produce particularly strong luminescence in the visible range, while luminescence in direct bandgap semiconductors corresponds to the bandgap

Fig. 4.3 Composite image of the luminescence traces inside a sapphire crystal versus the input pulse energy, as induced by filamentation of 210 fs, 1.3 μm laser pulse with a 23 μm FWHM spot size of the input beam. The laser beam propagates from left to right; $z = 0$ mm and $z = 4$ mm correspond to the input and output faces of the crystal, respectively

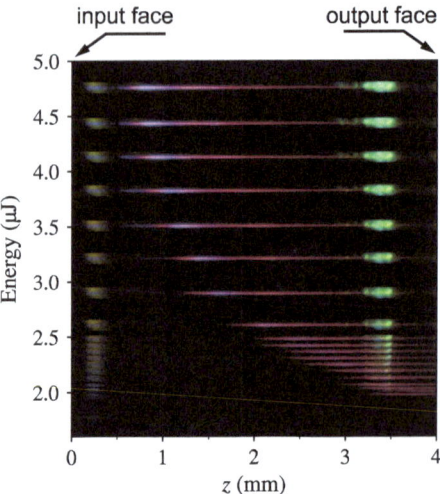

edge emission. Whatever is the mechanism of filament-induced luminescence, it readily serves to map the intensity variation within the light filaments, allowing the observer to monitor the filament formation dynamics, and to estimate a number of relevant parameters, such as filament diameter, peak intensity, and free electron density [13]. An example of filament-induced luminescence traces in a sapphire crystal is presented in Fig. 4.3. The most intense part of the luminescence trace indicates the highest intensity and hence the position of the nonlinear focus: note how the nonlinear focus moves from the output face toward the input face of the crystal as the input pulse energy increases. The intense green spots and more faint yellowish spots close to the output and input faces of the crystal, respectively, are parasitic reflections of the SC light. See also that no luminescence is produced for the input pulse energies below ∼2 μJ, which is the threshold energy for filament formation under given operating conditions.

In a practical setup, a variable density filter for fine adjustment of the pump pulse energy is more preferable than a hard iris aperture, since in the latter case the energy adjustment changes the NA of the input beam, thus altering the focusing condition. Moreover, self-focusing dynamics of clean and truncated Gaussian beams are somewhat different; truncated Gaussian beam is more sensitive to modulational instability, which leads to multiple filamentation if the input pulse power is sufficiently high. Slightly converging or diverging laser beams may be also in use; then the position of the nonlinear focus is defined by

$$\frac{1}{z'_{sf}} = \frac{1}{z_{sf}} + \frac{1}{f}, \tag{4.2}$$

Fig. 4.4 Red-shifted portions of SC spectra in sapphire as functions of the numerical aperture. Adapted from [15] and reprinted by permission from Springer Nature

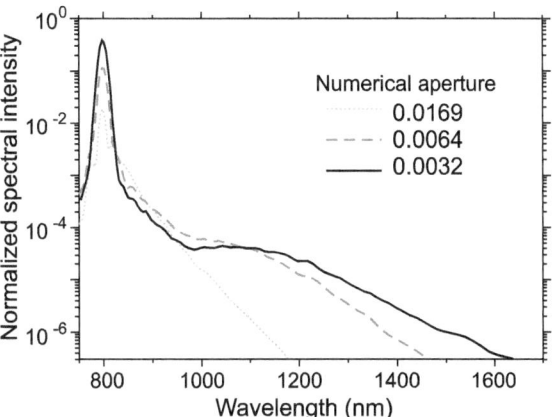

where f is the focal length of the focusing lens. Note that with converging input beams, the nonlinear focus occurs before the geometrical focus. In the case of diverging input beams (the geometrical focus is located before the input face of the nonlinear medium), the input pulse power should exceed P_{cr} by several times, since the self-focusing effect should overcome the divergence of the beam. A recent study of SC generation in water with converging and diverging input beams demonstrated that the most favorable conditions for an efficient SC generation are achieved when the focal plane of the input beam is located slightly before the input face of the nonlinear medium [14]. In the case of low NA, in such focusing condition, a slightly diverging input beam produces the SC with higher spectral energy density. On the other hand, experimental observations suggest that by setting the geometrical focus of the input beam inside the nonlinear medium, the energy threshold for SC generation could be reduced by as much as 25%, compared with beam focusing onto the entrance face of the sample.

Experiments also show that in a given normally dispersive nonlinear medium, a considerable enhancement of the red-shifted broadening of the SC spectrum may be achieved in loose focusing condition (NA < 0.01) [15, 16]. Figure 4.4 shows an example of this phenomenon, presenting experimentally measured red-shifted portions of SC spectra generated with 800 nm, 120 fs laser pulses in a sapphire crystal under various NA. Numerical simulations revealed that enhanced red-shifted spectral broadening stems from the increased nonlinear propagation of the leading sub-pulse, which emerges after the pulse spitting event and which preserves a steep ascending front over a larger propagation distance than in the case of high NA focusing conditions [15]. Interestingly, the blue-shifted spectral broadening is generally independent on the external focusing condition and its short-wavelength cutoff remains fairly constant.

4.3   Stability Issues

Pulse-to-pulse fluctuations of the SC spectral intensity and long-term stability are central aspects concerning many practical applications of the SC. The short and long-term instabilities originate from rather different factors.

It is obvious that the stability of the incident pump pulse energy determines the stability of the spectral components comprising the entire SC spectrum. It is widely considered that under optimized pumping conditions, fluctuations of the SC spectral intensity are very close to those of the pump pulse. Modern solid-state lasers provide excellent pulse-to-pulse energy stability under long-term operation, with the root mean square (RMS) fluctuations typically well below 1%, and so similar stability of the SC radiation is achieved, see e.g., [17]. However, the experimental analysis has shown that energy fluctuations of SC spectral components within selected narrow spectral intervals have distinct minima that occur for different pump energies [16], as illustrated in Fig. 4.5. These findings were confirmed by a more detailed statistical study performed over the entire SC wavelength range, that captured multiple stability and instability regions within the entire SC spectrum [18], as illustrated in Fig. 4.6. The exact locations of these specific stability/instability regions depend on the pump pulse energy; the existence of these regions is the signature of nonoptimal pump energy. The instabilities become almost totally suppressed just as the optimal pump pulse energy is set; usually, the optimal pump energy is justified by the saturation of the blue-shifted spectral broadening.

More rigorous statistical analysis revealed that under nonoptimal pumping conditions, e.g., slightly lower pump energy, which produces a slightly narrower SC spectrum and which could be regarded as a transient stage of SC generation, the intensities of the blue-shifted spectral components exhibit very large excursions from the average values obeying non-Gaussian, the so-called extreme-value (or rogue wave) statistics, producing highly asymmetric statistical distributions with well-pronounced

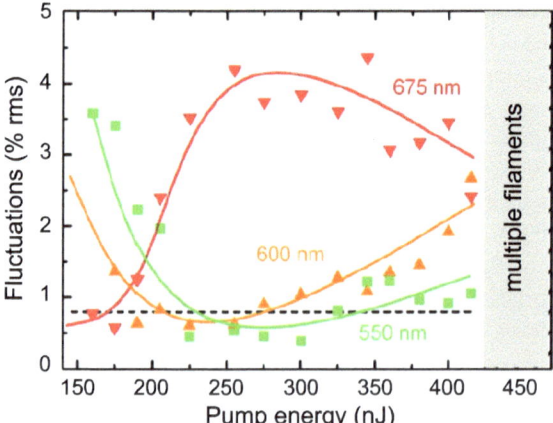

Fig. 4.5 Fluctuations of selected spectral components of the SC generated by 150 fs, 775 nm laser pulses in a 4 mm YAG crystal versus the pump pulse energy. The dashed line marks the level of energy fluctuations of the pump pulse. Reprinted from [16] by permission from Springer Nature

Fig. 4.6 **a** SC spectra in
sapphire as produced by 120
fs, 800 nm laser pulses
(NA = 0.01) with energies of
0.45 µJ (dotted curve) and
0.75 µJ (solid curve). **b**
Corresponding fluctuations
of the SC spectral intensity.
RMS fluctuations of the
input pulse are 0.4%, whose
level is indicated by a dashed
line at the bottom of the plot.
Adapted from the results of
[18] and reprinted by
permission from the Optical
Society of America

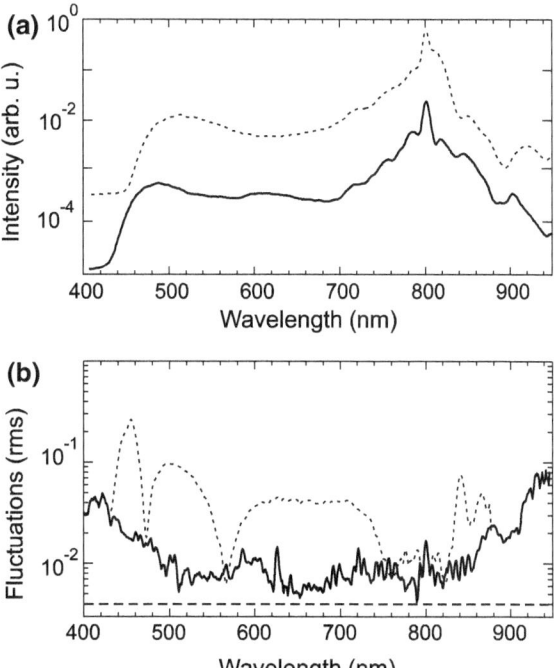

long tails [19]. In other words, such statistical distribution suggests that events with very high spectral intensity occur much more frequently than could be expected from normal-value (Gaussian) statistics. These peculiar statistical features originate from the intensity and phase noise of the pump pulses, which are inevitably present and hardly controlled in real experiments. The optimum pumping conditions imply that these small fluctuations are efficiently suppressed by the intensity clamping effect, and the resulting intensity fluctuations of the SC spectral components obey Gaussian statistics over the entire wavelength range.

Several statistical studies have disclosed that spectral components of the SC are correlated spectrally, temporally, and spatially, and these correlations exhibit complex evolution as a function of the pump pulse energy [18, 20, 21]. Although physical effects governing the observed complex correlation dynamics are not completely understood, spectral correlations are qualitatively explained by four-wave mixing [18], temporal correlations are attributed to abrupt processes of self-steepening [20], while spatial correlations emerge as a result of spectro-spatial couplings in the SC beam, which evolve in the filamentation process [21].

Long-term stability and reproducibility of the SC spectra are related mostly to the optical degradation of the material (formation of color centers, permanent modification of the refractive index, and eventually, the optical damage due to heat accumulation), which develop under long-term irradiation by repetitive laser pulses. Typically, with femtosecond pump pulses at 1 kHz repetition rate, the widely used sapphire and

YAG crystals operate well in the static setup, whereas fluoride crystals, such as CaF_2, BaF_2, MgF_2, and LiF, which provide the SC spectra with the largest ultraviolet extensions, require continuous translation or rotation [22]. The problem of the optical degradation becomes more severe for pumping with either long femtosecond, sub-picosecond and picosecond laser pulses, or at very high (from multi-kHz to MHz) repetition rates. So far, only YAG crystal demonstrated reliable operation under these adverse pumping conditions.

4.4 Effect of Filament Refocusing

Under loose focusing conditions (for numerical apertures below 0.05), and for sufficiently long nonlinear media and input beam powers exceeding P_{cr} by several times, a femtosecond filament may undergo recurrent self-focusing cycles. The general interpretation of self-focusing/refocusing cycles is based on the so-called dynamic spatial replenishment scenario, which assumes alternating cycles of self-focusing due to the Kerr effect and self-defocusing due to the free-electron plasma and which was originally proposed as an alternative to the self-guiding scenario to explain long-distance propagation of high power pulses in gaseous media [23]. A more recent experimental and numerical study captured the entire spatiotemporal evolution of light filaments versus the propagation distance in water and unveiled the intimate connections between complex propagation effects: focusing and refocusing cycles, nonlinear absorption, pulse splitting and replenishment, supercontinuum generation, and conical emission [24]. More specifically, whenever a self-focusing ultrashort-pulsed light beam approaches the nonlinear focus, multiphoton absorption attenuates its central part and induces pulse splitting and reshaping into a ring-like structure, as illustrated in Fig. 4.7a. With further propagation, the leading and trailing sub-pulses having slightly different carrier frequencies depart from each other due to material GVD, whereas the light contained in the ring replenishes the pulse on the propagation axis, as shown in Fig. 4.7a. If the power of the replenished pulse is above critical, the replenished pulse undergoes another self-focusing cycle, which results in the pulse splitting at second nonlinear focus, as illustrated in Fig. 4.7b. The second pulse-splitting produces yet another portion of the SC, and the resulting SC spectrum as well as the pattern of conical emission develop a periodic modulation, due to interference between split sub-pulses originating from the first and the second splitting events. After the second splitting event, pulse replenishment, refocusing and splitting may repeat once again, producing a third splitting event, after which the modulation in the SC spectrum has beatings contributed by the occurrence of a tertiary split sub-pulses. The refocusing cycles may continue as long as the power of the replenished pulse is still above critical. However, each refocusing cycle is followed by a sudden decrease in the transmittance due to multiphoton absorption and therefore the entire beam will continuously lose energy during propagation, and eventually, a linear propagation regime is resumed.

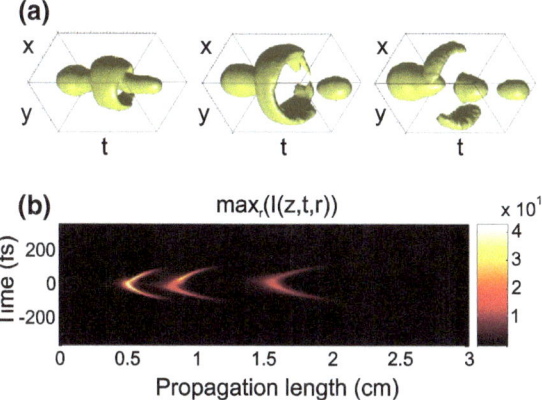

Fig. 4.7 **a** Experimentally measured spatiotemporal intensity profiles $I(x, y, t)$ of an ultraviolet filament in water at various stages of propagation after the first nonlinear focus, demonstrating pulse splitting and replenishment. **b** Numerical simulation of temporal profiles for the axial intensity distribution illustrating recurrent pulse splitting events associated with the nonlinear foci of a 90 fs ultraviolet (400 nm) laser pulse with 400 nJ energy propagating in a water cell. Adapted from [5] with permission

Focusing/refocusing cycles could be readily visualized by monitoring the intensity variation of the filament-induced luminescence traces from a side view of the nonlinear medium, see e.g., [25–27]. Figure 4.8 illustrates the above considerations by comparing visually perceptible filamentation features and SC spectra in a YAG crystal, as generated with 100 fs, 800 nm input pulses with energies of 310 nJ (peak power of 3.6 P_{cr}) and 560 nJ (6.6 P_{cr}), that induce a single self-focusing event and refocusing of the filament, respectively. The single self-focusing event is visualized by a gradually decaying filament-induced luminescence trace, whose most intense part indicates the position of the nonlinear focus, as shown in Fig. 4.8a. In this case, a featureless far-field pattern of SC emission (Fig. 4.8c) and a smooth SC spectrum (Fig. 4.8e) are produced. In contrast, the refocusing of the filament produces a double-peaked luminescence trace, as shown in Fig. 4.8b, and results in the occurrence of modulation in the outer part of the far-field pattern of SC emission (Fig. 4.8d) and periodic modulation of the SC spectrum (Fig. 4.8f). These indications are very important from a practical point of view, since they allow us to easily identify the recurrent collapse and pulse splitting events and optimize the operating conditions for SC generation without employing complex experimental measurements.

Despite the differences of temporal dynamics in the filamentation regime in the range of anomalous GVD, where pulse self-compression rather than splitting at the nonlinear focus takes place, recurrent self-focusing cycles manifest themselves in a similar manner [28, 29]. The refocusing event was shown to produce splitting of the light bullet at the secondary nonlinear focus [30], while multiple refocusing cycles were demonstrated to yield a sequence of quasi-periodic light bullets, each of them resulting in the ejection of a new portion of the SC. However, the above

Fig. 4.8 Luminescence traces in a YAG crystal produced by filamentation of 100 fs, 800 nm input pulses with energies of **a** 310 nJ, **b** 560 nJ, that represent a single self-focusing event and refocusing of the filament, respectively. **c** and **d** show the corresponding far-field patterns of SC emission, **e** and **f** show the corresponding SC spectra. Dashed curves show the spectrum of the input pulse. See text for details. Reprinted from [5] with permission

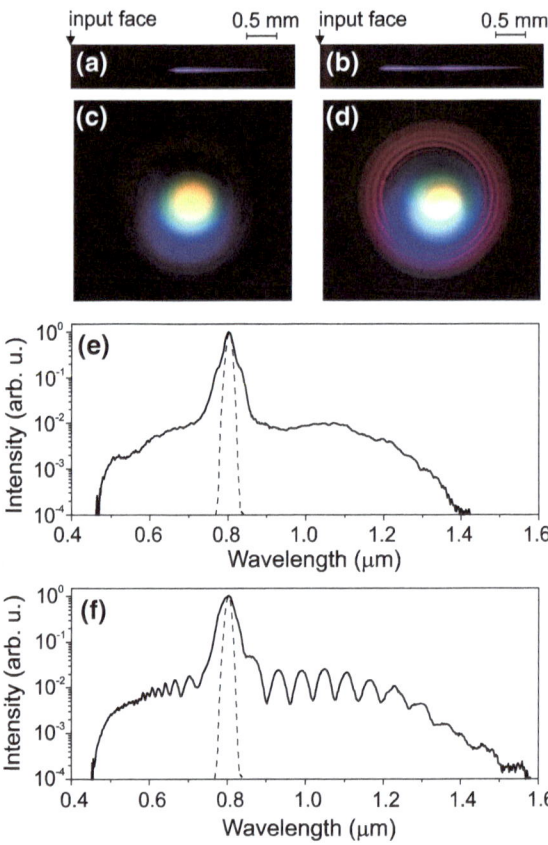

relationship between the filament refocusing and spectral broadening is truly valid for filamentation of ~100 fs and shorter input pulses, but appears qualitatively different in the case of SC generation with relatively long femtosecond, sub-picosecond and picosecond input pulses, as will be outlined in Sect. 6.3.

4.5 Multiple Filamentation

In the case of low numerical aperture, further increase of the pump pulse energy typically leads to beam breakup into several or multiple filaments, which emerge due to modulational instability that facilitates small-scale self-focusing of different parts of the input beam. Although multiple filamentation creates an illusion of SC energy scaling, beam breakup into multiple filaments results in deterioration of the spatial uniformity of the output beam and the temporal structure of the pulse, and eventually induces a considerable depolarization at various parts of the SC spectrum

[31, 32]. From a general point of view, the spectral extent of each individual filament is defined by a combined effect of the intensity clamping and the dispersion landscape of the given nonlinear material, so no additional spectral broadening by increasing the input beam power (the number of filaments) could be expected.

Multiple filaments may emerge in the form of either regular or irregular patterns, whose spatial intensity distributions are governed by the input beam size, symmetry, and smoothness. In particular, multiple filamentation induced by self-focusing of circular beams leads to non-reproducible irregular spatial distributions of the individual filaments, whose number and locations vary from one laser shot to another, see e.g., [33]. In contrast, self-focusing of elliptic laser beams produces regular and reproducible filament distributions in space: in the case of small input beam ellipticity the emerging filaments reside on the major and minor axes of an ellipse [34], while self-focusing of highly elliptic laser beams (e.g., as focused by a cylindrical lens) produces filament arrays with almost regular spacing between the adjacent filaments, each of them producing the SC [35]. Neighboring filaments exhibit mutual coherence, and this feature is exploited for the production of arrays of coherent broadband sources [36, 37]. Regular and well-reproducible multiple filamentation patterns were also produced by meshes and grids superimposing a regular diffraction structure on the input beams that act as seed cells dominating over random fluctuations in the self-focusing process in various liquid and solid-state media [38–40]. Controlled manipulation of the multifilamentation pattern demonstrating high- power SC generation without damaging the nonlinear medium was also reported using a microlens array [41, 42].

References

1. Brodeur, A., Chin, S.L.: Band-gap dependence of the ultrafast white-light continuum. Phys. Rev. Lett. **80**, 4406–4409 (1998)
2. Brodeur, A., Chin, S.L.: Ultrafast white-light continuum generation and self-focusing in transparent condensed media. J. Opt. Soc. Am. B **16**, 637–650 (1999)
3. Kolesik, M., Katona, G., Moloney, J.V., Wright, E.M.: Physical factors limiting the spectral extent and band gap dependence of supercontinuum generation. Phys. Rev. Lett. **91**, 043905 (2003)
4. Frolov, S.A., Trunov, V.I., Leshchenko, V.E., Pestryakov, E.V.: Multi-octave supercontinuum generation with IR radiation filamentation in transparent solid-state media. Appl. Phys. B **122**, 124 (2016)
5. Dubietis, A., Tamošauskas, G., Šuminas, R., Jukna, V., Couairon, A.: Ultrafast supercontinuum generation in bulk condensed media (Review). Lith. J. Phys. **57**, 113–157 (2017)
6. Gallais, L., Commandré, M.: Laser-induced damage thresholds of bulk and coating optical materials at 1030 nm, 500 fs. Appl. Opt. **53**, A186–A196 (2014)
7. Mero, M., Liu, J., Rudolph, W., Ristau, D., Starke, K.: Scaling laws of femtosecond laser pulse induced breakdown in oxide films. Phys. Rev. B **71**, 115109 (2005)
8. Gallais, L., Douti, D.-B., Commandré, M., Batavičiūtė, G., Pupka, E., Ščiuka, M., Smalakys, L., Sirutkaitis, V., Melninkaitis, A.: Wavelength dependence of femtosecond laser-induced damage threshold of optical materials. J. Appl. Phys. **117**, 223103 (2015)

9. Ashcom, J.B., Gattass, R.R., Schaffer, C.B., Mazur, E.: Numerical aperture dependence of damage and supercontinuum generation from femtosecond laser pulses in bulk fused silica. J. Opt. Soc. Am. B **23**, 2317–2322 (2006)
10. Nguyen, N.T., Saliminia, A., Liu, W., Chin, S.L., Valée, R.: Optical breakdown versus filamentation in fused silica by use of femtosecond infrared laser pulses. Opt. Lett. **28**, 1591–1593 (2003)
11. Kudarauskas, D., Tamošauskas, G., Vengris, M., Dubietis, A.: Filament-induced luminescence and supercontinuum generation in undoped, Yb-doped and Nd-doped YAG crystals. Appl. Phys. Lett. **112**, 041103 (2018)
12. Jukna, V., Garejev, N., Tamošauskas, G., Dubietis, A.: Role of external focusing geometry in supercontinuum generation in bulk solid-state media. J. Opt. Soc. Am. B **36**, A54–A60 (2019)
13. Dharmadhikari, A.K., Rajgara, F.A., Mathur, D.: Plasma effects and the modulation of white light spectra in the propagation of ultrashort, high-power laser pulses in barium fluoride. Appl. Phys. B **82**, 575–583 (2006)
14. Potemkin, F.V., Mareev, E.I., Smetanina, E.O.: Influence of wavefront curvature on supercontinuum energy during filamentation of femtosecond laser pulses in water. Phys. Rev. A **97**, 033801 (2018)
15. Jukna, V., Galinis, J., Tamošauskas, G., Majus, D., Dubietis, A.: Infrared extension of femtosecond supercontinuum generated by filamentation in solid-state media. Appl. Phys. B **116**, 477–483 (2014)
16. Bradler, M., Baum, P., Riedle, E.: Femtosecond continuum generation in bulk laser host materials with sub-µJ pump pulses. Appl. Phys. B **97**, 561–574 (2009)
17. Megerle, U., Pugliesi, I., Schriever, C., Sailer, C.F., Riedle, E.: Sub-50 fs broadband absorption spectroscopy with tunable excitation: putting the analysis of ultrafast molecular dynamics on solid ground. Appl. Phys. B **96**, 215–231 (2009)
18. Majus, D., Dubietis, A.: Statistical properties of ultrafast supercontinuum generated by femtosecond Gaussian and Bessel beams: a comparative study. J. Opt. Soc. Am. B **30**, 994–999 (2013)
19. Majus, D., Jukna, V., Pileckis, E., Valiulis, G., Dubietis, A.: Rogue-wave-like statistics in ultrafast white-light continuum generation in sapphire. Opt. Express **19**, 16317–16323 (2011)
20. Bradler, M., Riedle, E.: Temporal and spectral correlations in bulk continua and improved use in transient spectroscopy. J. Opt. Soc. Am. B **31**, 1465–1475 (2014)
21. van de Walle, A., Hanna, M., Guichard, F., Zaouter, Y., Thai, A., Forget, N., Georges, P.: Spectral and spatial full-bandwidth correlation analysis of bulk-generated supercontinuum in the mid-infrared. Opt. Lett. **40**, 673–675 (2015)
22. Wang, J., Zhang, Y., Shen, H., Jiang, Y., Wang, Z.: Spectral stability of supercontinuum generation in condensed mediums. Opt. Eng. **56**, 076107 (2017)
23. Mlejnek, M., Wright, E.M., Moloney, J.V.: Dynamic spatial replenishment of femtosecond pulses propagating in air. Opt. Lett. **23**, 382–384 (1998)
24. Jarnac, A., Tamošauskas, G., Majus, D., Houard, A., Mysyrowicz, A., Couairon, A., Dubietis, A.: Whole life cycle of femtosecond ultraviolet filaments in water. Phys. Rev. A **89**, 033809 (2014)
25. Wu, Z.X., Jiang, H.B., Luo, L., Guo, H.C., Yang, H., Gong, Q.H.: Multiple foci and a long filament observed with focused femtosecond pulse propagation in fused silica. Opt. Lett. **27**, 448–450 (2002)
26. Liu, W., Chin, S.L., Kosareva, O., Golubtsov, I.S., Kandidov, V.P.: Multiple refocusing of a femtosecond laser pulse in a dispersive liquid (methanol). Opt. Commun. **225**, 193–209 (2003)
27. Dharmadhikari, A.K., Dharmadhikari, J.A., Mathur, D.: Visualization of focusing-refocusing cycles during filamentation in BaF$_2$. Appl. Phys. B **94**, 259–263 (2009)
28. Chekalin, S.V., Dokukina, A.E., Dormidonov, A.E., Kompanets, V.O., Smetanina, E.O., Kandidov, V.P.: Light bullets from a femtosecond filament. J. Phys. B **48**, 094008 (2015)
29. Kuznetsov, A.V., Kompanets, V.O., Dormidonov, A.E., Chekalin, S.V., Shlenov, S.A., Kandidov, V.P.: Periodic colour-centre structure formed under filamentation of mid-IR femtosecond laser radiation in a LiF crystal. Quantum Electron. **46**, 379–386 (2016)

30. Gražulevičiūtė, I., Šuminas, R., Tamošauskas, G., Couairon, A., Dubietis, A.: Carrier-envelope phase-stable spatiotemporal light bullets. Opt. Lett. **40**, 3719–3722 (2015)
31. Dharmadhikari, A.K., Rajgara, F.A., Mathur, D.: Depolarization of white light generated by ultrashort laser pulses in optical media. Opt. Lett. **31**, 2184–2186 (2006)
32. Kumar, R.S.S., Deepak, K.L.N., Rao, D.N.: Depolarization properties of the femtosecond supercontinuum generated in condensed media. Phys. Rev. A **78**, 043818 (2008)
33. Bergé, L., Mauger, S., Skupin, S.: Multifilamentation of powerful optical pulses in silica. Phys. Rev. A **81**, 013817 (2010)
34. Dubietis, A., Tamošauskas, G., Fibich, G., Ilan, B.: Multiple filamentation induced by input-beam ellipticity. Opt. Lett. **29**, 1126–1128 (2004)
35. Majus, D., Jukna, V., Valiulis, G., Dubietis, A.: Generation of periodic filament arrays by self-focusing of highly elliptical ultrashort pulsed laser beams. Phys. Rev. A **79**, 033843 (2009)
36. Cook, K., Kar, A.K., Lamb, R.A.: White-light supercontinuum interference of self-focused filaments in water. Appl. Phys. Lett. **83**, 3861–3863 (2003)
37. Corsi, C., Tortora, A., Bellini, M.: Generation of a variable linear array of phase-coherent supercontinuum sources. Appl. Phys. B **78**, 299–304 (2004)
38. Schroeder, H., Liu, J., Chin, S.L.: From random to controlled small-scale filamentation in water. Opt. Express **12**, 4768–4774 (2004)
39. Liu, L., Schroeder, H., Chin, S.L., Li, R., Xu, Z.: Ultrafast control of multiple filamentation by ultrafast laser pulses. Appl. Phys. Lett. **87**, 161105 (2005)
40. Dharmadhikari, A.K., Rajgara, F.A., Mathur, D., Schroeder, H., Liu, J.: Efficient broadband emission from condensed media irradiated by low-intensity, unfocused, ultrashort laser light. Opt. Express **13**, 8555–8564 (2005)
41. Cook, K., McGeorge, R., Kar, A.K., Taghizadeh, M.R., Lamb, R.A.: Coherent array of white-light continuum filaments produced by diffractive microlenses. Appl. Phys. Lett. **86**, 021105 (2005)
42. Camino, A., Hao, Z., Liu, X., Lin, J.: High spectral power femtosecond supercontinuum source by use of microlens array. Opt. Lett. **39**, 747–750 (2014)

Chapter 5
Experimental Results

This chapter presents a comprehensive overview of the literature and up-to-date experimental results on supercontinuum generation in commonly used wide-bandgap dielectric materials: water, fused silica, alkali metal fluorides, and laser host crystals, achieved so far with a wide variety of pump wavelengths, ranging from the ultraviolet to the mid-infrared, accessing the filamentation regimes under normal, zero, and anomalous GVD in these materials. Alongside these results, the state of the art of supercontinuum generation is presented for a number of less explored, but potentially useful nonlinear materials: non-silica glasses, crystals with second-order nonlinearity, semiconductors, narrow bandgap dielectric, and others, which are advantageous for SC generation in the mid-infrared.

5.1 Water as a Prototypical Nonlinear Medium

Due to the absence of permanent optical damage and the possibility to easily vary the medium thickness, liquids are regarded as attractive materials for many experiments in ultrafast light–matter interactions and nonlinear optics in particular. Water occupies an exceptional place among the other liquids because of its technological and biomedical importance and therefore often serves as a prototypical nonlinear medium for studies of femtosecond filamentation and SC generation. Systematic studies of the spectral broadening in water date back to the mid-1980s [1] and successfully continue in the femtosecond laser era, see [2–5] for early accounts on femtosecond SC generation in water and other liquids.

Generation of femtosecond SC in water was studied with pump wavelengths, which cover almost its entire transparency range. With ultraviolet pumping, the measured SC spectra in water covered the wavelength ranges of 290–530 nm [6] and 350–550 nm [7], as reported with sub-100 fs pulses with wavelengths of 393 nm and 400 nm, produced by frequency doubling the output of an amplified Ti:sapphire

© The Author(s), under exclusive license to Springer Nature Switzerland AG 2019
A. Dubietis and A. Couairon, *Ultrafast Supercontinuum Generation
in Transparent Solid-State Media*, SpringerBriefs in Physics,
https://doi.org/10.1007/978-3-030-14995-6_5

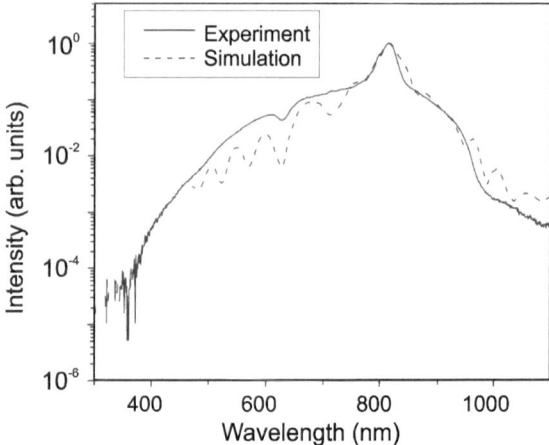

Fig. 5.1 Typical supercontinuum spectrum in water, as generated by filamentation of 45 fs, 810 nm Ti:sapphire laser pulses with an energy of 3 μJ in 2 cm-long water cell. Adapted from [10]. Reprinted by permission from Springer Nature

laser, respectively, and under slightly different external focusing conditions. When pumped in the visible spectral range, the SC spectrum from 400 to 650 nm was generated with 527 nm self-compressed femtosecond second harmonic pulses from a Nd:glass laser [8] and from 450 to 720 nm with 594 nm pulses from a rhodamine 6G dye laser [5]. A considerable effort was dedicated to study SC generation in the near infrared under various operating conditions and external focusing geometries, using the fundamental wavelength (around 800 nm) of Ti:sapphire lasers [9–13]. A typical SC spectrum in water that covers the 400–1100 nm spectral range is illustrated in Fig. 5.1. A broader SC spectrum spanning wavelengths from 350 to 1400 nm was reported using a thin water jet instead of a thick cell [14].

In the range of anomalous GVD of water (for pump wavelengths longer than 1 μm), using 200 fs pump pulses with a central wavelength of 1.24 μm from a Cr:forsterite laser system, the SC spectrum extending from 410 nm to above 1.6 μm was measured under pumping geometry that was optimized for achieving the most stable SC output, by using a slightly diverging pump beam [15]. Although rapidly increasing infrared absorption due to the presence of strong absorption bands located at 1.46 and 1.94 μm is considered as the main factor that limits the redshifted broadening of the SC spectrum in water, an unexpectedly broad SC spectrum, ranging from 350 nm to 1.75 μm was reported using 1.3 μm pump pulses from an optical parametric amplifier [16]. A more recent study demonstrated that under carefully chosen experimental conditions, the absorption effects can be overcome using a thin water jet, thus reducing the nonlinear interaction length just to a few tens of micrometers [17]. Under these experimental settings, with the pump wavelength set at 1.6 μm, a SC spectrum was produced, spanning more than two octaves from 300 nm to 2.4 μm.

Table 5.1 Relevant linear and nonlinear properties of glasses used for SC generation: the energy bandgap (U_g), transmittance, the nonlinear refractive index (n_2), and the zero GVD wavelength (λ_0)

Material	U_g eV	Transmittance μm	n_2 $\times 10^{-16}$ cm^2/W	λ_0 μm
Fused silica	9.0	0.18–3.5	2.4	1.27
ZBLAN	6.2	0.25–6.9	2.9	1.7
BK7	4.28	0.3–2.5	3.75	1.32
La glass	4.1	0.3–3.6	11.4	1.8
Te glass	3.5	0.38–6.15	25	1.95
As$_2$S$_3$	2.5	0.59–13	250	4.8

5.2 Glasses

Glasses constitute a large family of multifunctional optical materials, which find diverse applications in contemporary optical sciences and technology. Table 5.1 provides relevant optical characteristics of commonly used glasses for SC generation.

Fused silica (SiO$_2$) is often regarded as an etalon solid-state nonlinear medium thanks to a large energy bandgap, a reasonably large nonlinear index of refraction, high optical and mechanical quality, combined with precise knowledge of relevant material parameters. At present, fused silica is often used in practical schemes for pulse compression that involve self-phase modulation-induced spectral broadening, see Sect. 6.2 for more details.

Spectral broadening and SC generation in fused silica was investigated using a wide variety of pump wavelengths, ranging from the deep ultraviolet to the mid-infrared. Only very slight spectral broadening around the carrier wavelength was observed in the deep ultraviolet with femtosecond pump pulses at 248 nm from an amplified excimer laser [18]. Spectral broadening to a similar extent was measured using pump pulses at 262 nm, produced by frequency tripling of a Ti:sapphire laser output [6]. More noticeable, but still modest spectral broadening in fused silica and BK7 glass was reported with the second harmonic pulses (393 nm) of the Ti:sapphire laser [6]. With a pump wavelength located in the visible spectral range, a SC spectrum spanning wavelengths from 415 to 720 nm was generated with 250 fs pulses at 594 nm from a rhodamine 6G dye laser [5].

A large number of experiments on SC generation in fused silica were performed with near-infrared pumping, using the fundamental harmonic of a Ti:sapphire laser with input pulsewidths of 100 fs and shorter. A typical SC spectrum in fused silica produced with ∼800 nm pulses extends from 390 to 1000 nm, and is almost independent on the operating conditions (pulsewidth, energy, numerical aperture and sample length used) [6, 19–24]. Somewhat narrower SC spectra were reported in BK7 [6, 25, 26] and ZK7 [27] glasses, which possess smaller bandgaps. On the other hand, considerably broader SC spectra in silica glasses were produced by setting the

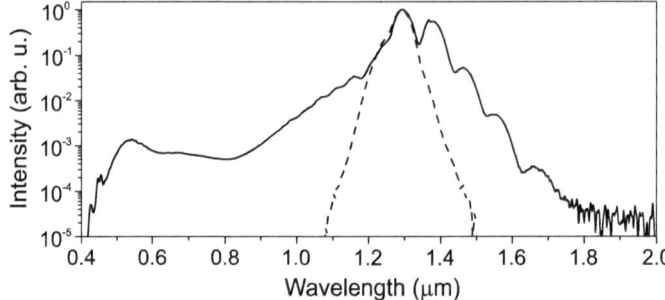

Fig. 5.2 Supercontinuum spectrum in a 3-mm-long UV-grade fused silica sample pumped near its zero GVD point by 1.3 μm, 100 fs, 1.9 μJ pulses from an optical parametric amplifier. The dashed curve shows the input pulse spectrum. Reprinted from [29] with permission

pump wavelengths close to or slightly above the zero GVD wavelengths [28, 30, 31], which are located around 1.3 μm, see Table 5.1. Figure 5.2 shows an example of the SC spectrum generated by 100 fs, 1.3 μm pump pulses in a 3-mm-thick UV-grade fused silica sample.

A couple of SC generation experiments were performed with pump pulses whose wavelengths fall into the range of anomalous GVD of fused silica, and uncovered a number of universal features, which characterize the entire shape of the SC spectrum in the spectral angular domain [32–34]. The angle-integrated as well as the axial SC spectra show an intense peak located in the visible range (the so-called blue peak), which is identified as an axial component of the conical emission and whose blueshift increases with increasing the wavelength of the driving pulse [34, 35]. Various aspects of the spectral broadening and SC generation in fused silica, such as energy content, stability of the carrier envelope phase, etc., were studied in connection to the formation and propagation dynamics of self-compressed spatiotemporal light bullets, which emerge from favorable interplay between self-phase modulation and anomalous GVD during the filamentation process [36, 37]. Experiments on self-focusing and filamentation in fused silica of few optical cycle pulses with carrier wavelengths of 2 μm [38] and 2.2 μm [39] reported ultrabroad SC spectra starting from 400 and 370 nm on the short-wavelength side, respectively, and extending up to wavelengths greater than 2.5 μm. More recently, the recorded dynamics of spectral broadening in fused silica with a pump wavelength of 2.3 μm versus the input pulse energy revealed that under given experimental conditions, there exists an optimum pump pulse energy to obtain the broadest SC spectrum [40], as illustrated Fig. 5.3a. Figure 5.3b shows the broadest SC spectrum generated with the input pulse energy of 2.8 μJ, providing a continuous wavelength coverage from 310 nm to 3.75 μm, which converts to 3.6 optical octaves and whose redshift extends slightly beyond the infrared absorption edge of fused silica.

Various non-silica glasses currently receive increasing attention as very promising nonlinear media for SC generation, especially aiming at spectral broadening in the mid-infrared spectral range. Using pump pulses at 1.6 μm from an optical para-

Fig. 5.3 a Experimentally measured spectral broadening of 100 fs, 2.3 μm laser pulse in a 3-mm-thick fused silica sample versus input pulse energy. The spectral peak centered at 767 nm, which appears before the onset of supercontinuum generation, is the third harmonic. **b** The broadest supercontinuum spectrum generated with the input pulse energy of 2.8 μJ. Note the intense blue peak centered at 430 nm. The input pulse spectrum is shown by the dashed curve. Adapted from [40]. Reprinted by permission from the Optical Society of America

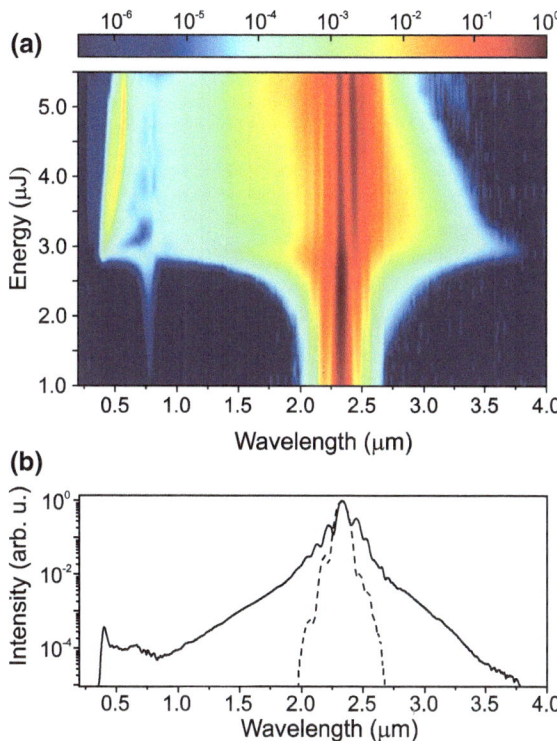

metric amplifier, the SC spectra spanning wavelengths from 400 to 2800 nm were reported in lanthanum glass [41]. An impressive spectral broadening was observed in fluoride glass (ZBLAN) under similar pumping conditions; the authors measured an ultrabroadband, more than 5 octaves-wide SC, covering almost the entire transmission range of the material (0.2–8.0 μm) [42]. The so-called soft glasses hold a great potential for SC generation in the mid-infrared spectral range. These materials are widely exploited in modern optical fiber technology, however, bulk soft glasses, such as tellurite and chalcogenide, also show very promising results. The SC spectra extending from the visible to 6 μm and from the visible to 4 μm were reported in tellurite glass using pump pulses with central wavelengths of 1.6 μm [43] and 2.05 μm [44], respectively. A spectrally flat SC in the 2.5–7.5 μm range was produced in bulk chalcogenide glass using pump pulses with a central wavelength of 5.3 μm [45]. SC spectra with remarkable mid-infrared coverage from 2.5 to ~11 μm were produced in As_2S_3 and GeS_3 bulk chalcogenide glass samples pumped by 65 fs pulses at 4.8 μm, that matched the zero GVD wavelength of As_2S_3 [46], see Fig. 5.4. Another study reported on the generation of a mid-infrared SC extending from 2.44 to 12 μm in chalcogenide glass samples of various composition, when pumped with 70 fs pulses with tunable carrier wavelength in the 3.75–5.0 μm range [47].

Fig. 5.4 Supercontinuum
spectra produced by
filamentation of 65 fs, 4.8
μm pulses in 3.3-mm-long
samples of As_2S_3 and GeS_3.
The inset shows measured
spectral transmission of the
samples. Reprinted from
[46] by permission from
Springer Nature

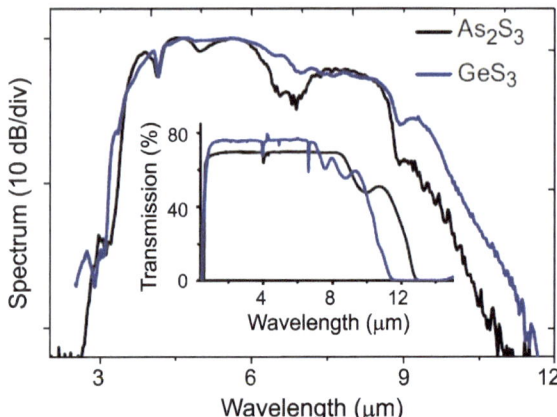

More recently, it was demonstrated that change of the composition of chalcogenide glasses by introducing more polarizable elements favorably modifies their optical properties by increasing nonlinearity and transmission windows and also shifting the corresponding zero dispersion wavelengths farther into the infrared region, making these materials well suited for broadband SC generation [48]. To this end, SC spectra spanning more than 2 octaves with the redshifted spectral broadening reaching into the far infrared were measured in GeSSe (in the 2.6–11 μm range), GeSe (2.6–12 μm), and TGG (2.5–16 μm) bulk glass samples of 4 mm thickness, when pumped by 65 fs pulses with carrier wavelengths of 4.5 μm (for GeSSe and GeSe) and 7.3 μm (for TGG).

5.3   Alkali Metal Fluorides

Most alkali metal fluorides possess the largest bandgaps among solid-state dielectric materials. Consequently, these media exhibit extremely broad transparency windows, which extend from the vacuum ultraviolet to the mid-infrared (see Table 4.1 in Sect. 4), making them very attractive nonlinear materials for SC generation as pumped with femtosecond laser pulses at various parts of the optical spectrum. In particular, lithium and calcium fluorides, LiF and CaF_2, are the only solid-state materials that are able to produce an appreciable spectral broadening in the deep ultraviolet, as demonstrated with third harmonic pulses from a Ti:sapphire laser serving as a pump [6, 49]. When pumped with the second harmonic pulses from the same laser, the measured SC spectra in these materials cover the ultraviolet and visible spectral ranges [6, 50, 52].

Extensive experimental studies of SC generation in alkali metal fluorides were performed with fundamental harmonic pulses of the Ti:sapphire laser, with an emphasis on the blueshifted (anti-Stokes) part of the SC spectrum, which is relevant for

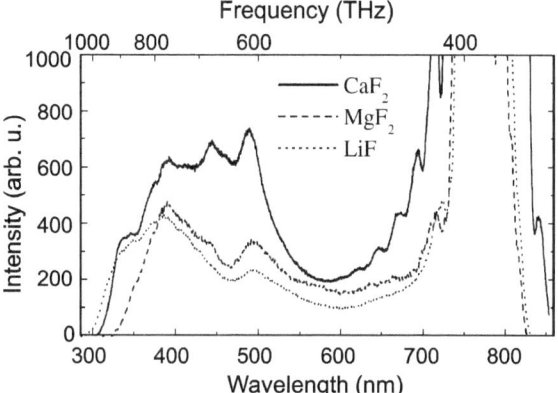

Fig. 5.5 Supercontinuum spectra in CaF$_2$, MgF$_2$ and LiF crystals, generated with 150 fs pump pulses at 775 nm. Adapted from [53]. Reprinted by permission from Elsevier

a number of applications. A comparative study of the SC generation in CaF$_2$, LiF and MgF$_2$ crystals demonstrated that LiF and CaF$_2$ crystals produce SC spectra with the largest frequency upshift on the ultraviolet side with cutoff wavelengths below 300 nm [53], as shown in Fig. 5.5. A similar trend regarding the blueshifted cutoff wavelengths was confirmed by a more recent study; however, slightly longer cutoff wavelengths were detected due to reduced dynamic range of the measurements [54].

Among the other alkali metal fluorides, CaF$_2$ is identified as the most useful material for stable SC generation in the ultraviolet and visible spectral range, as it exhibits low induced depolarization [51, 52, 55] and excellent reproducibility of the spectrum, which are important characteristics for femtosecond transient absorption spectroscopy, see, e.g., [49, 50, 56–58]. Moreover, owing to one of the largest ultraviolet extensions, the SC spectrum in CaF$_2$ was exploited for improvement of performance characteristics and extension of the tuning range of ultraviolet-pumped noncollinear optical parametric amplifiers [53, 59]. In the range of anomalous GVD of CaF$_2$, measurements performed with nearly two optical cycle (15 fs) pump pulses with a central wavelength at 2 μm demonstrated an exceptionally flat shape of the SC spectrum, yielding a broad plateau in the visible and near infrared, in the wavelength range of 500–1700 nm [38]. However, apart from the aforementioned advantages, CaF$_2$ has a relatively low optical damage threshold, if exposed to repetitive pulses, therefore a reliable and reproducible SC generation in this material is achieved only in the setups where continuous translation or rotation of the crystal is performed.

Barium fluoride, BaF$_2$, is a relatively inexpensive cubic crystal that serves as a general-purpose optical window material that is transparent from the ultraviolet to the long-wave infrared. BaF$_2$ is widely exploited nonlinear medium for SC generation, however, due to its slightly smaller bandgap compared to CaF$_2$, the SC cutoff wavelengths in BaF$_2$ are in the range of 320–350 nm, with pumping by fundamental harmonics of Ti:sapphire lasers [54, 60, 61]. On the other hand, BaF$_2$ is a well-known scintillator possessing two strong luminescence bands centered at 200 and 330 nm. The latter luminescence band was readily employed to map the intensity variation within light filaments; in particular, a side view of six-photon absorption-induced luminescence allowed monitoring the filament formation dynamics, focusing and refocusing cycles, and eventually estimating a number of relevant parameters, such as filament diameter, peak intensity, free electron density, and multiphoton absorption cross section [62, 63]. From a practical viewpoint, SC generation in BaF_2 was used to provide a broadband seed signal for OPCPA pumped by high repetition rate Yb:YAG thin disc regenerative amplifier; after amplification, the pulses were compressed down to 4.6 fs, which is very close to the Fourier limit of the amplified SC spectrum [64].

A number of experimental studies provide illustrative comparisons of the SC spectra generated in CaF_2 and BaF_2 crystals using pump wavelengths in the near- and mid-infrared spectral range. With pump pulses at 800 nm, SC spectra extending from 300 nm to 2 μm and from 320 nm to 1.98 μm were measured in CaF_2 and BaF_2 samples, respectively. An enhanced redshifted spectral broadening in these nonlinear crystals was recorded using pump pulses at 1.38 μm [39]. Using pump wavelengths of 2.1 and 2.2 μm, where both crystals feature anomalous GVD, combined data from [39, 65] yield ultrabroad, multioctave SC spectra in CaF_2 and BaF_2, spanning wavelengths from 340 nm to 3.3 μm and from 350 nm to 3.8 μm, respectively.

Experiments on SC generation in CaF_2 and BaF_2 with longer wavelength mid-infrared pulses (tunable in the 2500–3800 nm range) reported a dramatic change of the SC spectral shapes [66, 67]. These measurements showed that, for instance, with pump pulses at 3 μm, SC spectra are no longer continuous, but are composed of two separate bands, one located in the visible range (400–600 nm) and another in the mid-infrared (1500–4500 nm), while the spectral components in the near-infrared spectral range are practically absent or at least their spectral intensities fall below the detection range. More recent spectral measurements with 3.5 μm pulses (Fig. 5.6), performed over a higher dynamic range, revealed that the spectral discontinuity still exists in CaF_2, however, demonstrating more homogenous SC spectrum in BaF_2, which provided the spectral coverage from the visible to beyond 5 μm in the mid-infrared [68].

Although LiF possesses the largest (13.6 eV) bandgap among dielectric solid-state materials and therefore produces the largest blue shifts of the SC spectra, formation of persistent color centers in LiF is generally considered as a major drawback to its practical application for SC generation. However, recent experiments with Ti:sapphire laser pumping revealed that color centers only slightly modify the ultraviolet cutoff (around 270 nm) of the SC spectrum on the long-term (during several hours) operation [69].

An ultrabroad, almost 4 octave-spanning SC spectrum, continuously covering the wavelength range from 290 nm to 4.3 μm (at the 10^{-6} intensity level) was reported in LiF with 2.3 μm pump pulses, whose wavelength falls into range of anomalous GVD of the crystal [40]. Interestingly, such an ultrabroadband spectrum was recorded in the presence of color centers. The recorded evolution of the SC spectra in time revealed that spectral modifications due to the formation of color centers evolve on a very fast time scale (just a few tens of laser shots at a 1 kHz repetition rate). After a few thousands of laser shots, the SC spectrum eventually stabilizes and remains unchanged during further operation. Figure 5.7 compares the SC spectra generated

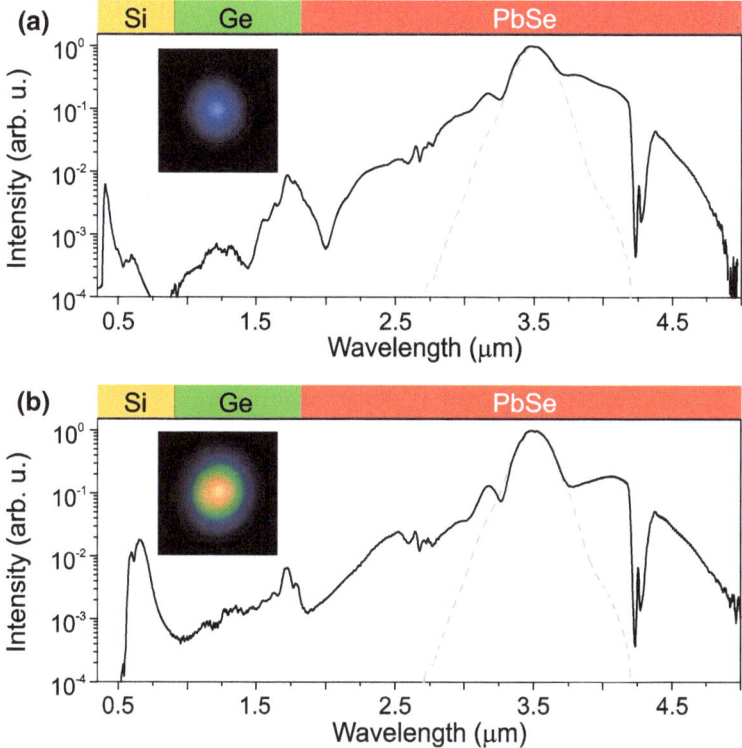

Fig. 5.6 Supercontinuum spectra generated with 60 fs, 31 µJ pulses at 3.5 µm in **a** CaF$_2$ and **b** BaF$_2$, both of 4 mm thickness. Dashed curves show the input pulse spectrum. A deep double dip around 4.25 µm is due to absorption of atmospheric CO$_2$. The ranges of spectrometer detectors (Si, Ge, PbSe) are indicated by color bars on the top. The insets show the visual appearances of the SC beams in the far field. Adapted from [68]. Reprinted by permission from IOP Publishing

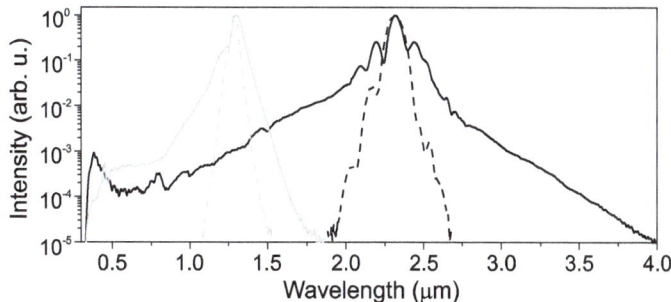

Fig. 5.7 A comparison of supercontinuum spectra in a 3.5-mm-thick LiF sample, produced by filamentation of 100 fs pulses with wavelengths of 1.3 µm (gray curve) and 2.3 µm (black curve), which fall into the ranges of zero and anomalous GVD of the crystal and with energies of 2.2 µJ and 9.5 µJ, respectively. Spectra of the input pulses are shown by dashed curves. Reprinted from [29] with permission

in a 3.5 mm-long LiF plate by 100 fs pulses with wavelengths of 1.3 and 2.3 μm, which fall into the ranges of zero and anomalous GVD of the crystal, respectively. The spectra exhibit a stable ultraviolet cutoff at 330 nm, as defined at the 10^{-5} intensity level. A series of experiments on SC generation in LiF with wavelength-tunable pump pulses from the optical parametric amplifier captured how the spectral position of the detached blue peak (that corresponds to an intensified visible SC band seen in Fig. 5.7) shifts from 270 to 500 nm while tuning the input wavelength from 3.3 to 1.9 μm [70, 71].

A more recent study has identified lead fluoride (PbF_2) as very promising non-linear material for SC generation owing to the relatively narrow bandgap ($U_g = 5$ eV), but very broad transparency window (0.29–12.5 μm) and large nonlinear index of refraction ($n_2 = 11.7 \times 10^{-16}$ cm^2/W) of this cubic crystal. The broadest SC spectrum spanning 4.7 octaves from 350 nm to 9 μm was produced using 180 fs, 1.6 μm pump pulses with the peak power greatly exceeding the critical power for self-focusing, yielding multiple filaments that induce persistent waveguides in the material [72].

5.4 Laser Hosts

Popular laser host crystals, such as undoped sapphire (Al_2O_3) and yttrium aluminum garnet ($Y_3Al_5O_{12}$, YAG) are excellent nonlinear media possessing high crystalline quality, relatively large nonlinearities, and high optical damage thresholds. Therefore, it is quite surprising that the potential of these nonlinear materials for SC generation was generally overlooked in the early systematic studies on this topic. Nevertheless, the first experimental demonstrations of SC generation in sapphire date back to 1994 [73, 74], revealing sapphire as a long sought solid-state material to replace at that time commonly used liquid media, putting the technology of femtosecond optical parametric amplifiers on all solid-state grounds, see also [75] for more details. Since then, sapphire became a routinely used nonlinear medium for SC generation with Ti:sapphire driving lasers, providing a high quality seed signal that is an indispensable asset in modern ultrafast optical parametric amplifiers, see, e.g., [76].

Although sapphire is rarely used for SC generation with pumping in the visible range, it produces an appreciable spectral broadening that is of importance for some applications. In that regard, a SC spectrum from 340 to 650 nm was generated using second harmonic pulses of an amplified Yb-fiber laser (515 nm) as a pump, which was thereafter used to seed a noncollinear optical parametric amplifier pumped by third harmonic (343 nm) pulses of the same laser, rendering a significantly enhanced wavelength tuning down to the near-ultraviolet spectral range [77].

Using pump pulses at the fundamental harmonic of a Ti:sapphire laser (pump wavelengths of around 800 nm) and under the commonly used operating conditions (input pulse energy of ~1 μJ and material thickness in the range of 1–3 mm), a typical SC spectrum produced in sapphire covers the wavelength range from 410 to 1100 nm [78–81]. A notable extension of the infrared part of the SC spectrum

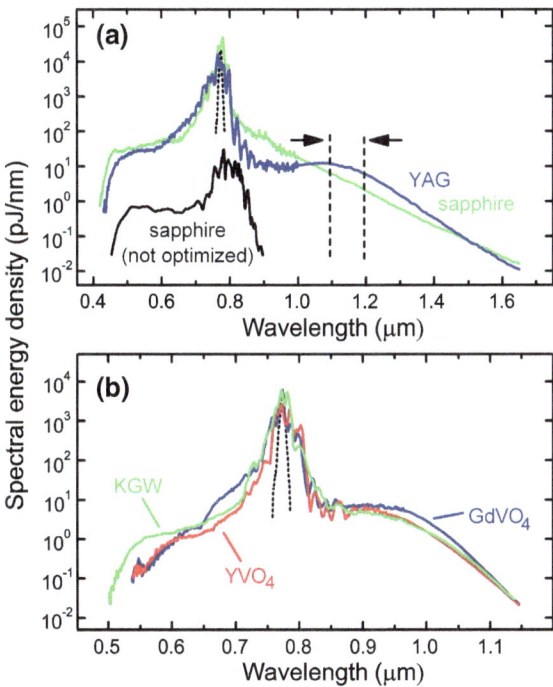

Fig. 5.8 a Black curve: supercontinuum spectrum from a 3-mm- thick sapphire sample with conventional (tight focusing) pumping conditions (shown not to scale, shifted for clarity). Green curve: supercontinuum spectrum from a 3-mm-thick sapphire sample with loose focusing condition. Blue curve: supercontinuum spectrum from a 4-mm-thick YAG sample showing an improved photon density in the infrared region. **b** Supercontinuum spectra generated in KGW (green curve), YVO$_4$ (red curve) and GdVO$_4$ (blue curve) crystals (all 4 mm thick) with the input pulse energies of 83 nJ, 59 nJ, and 78 nJ, respectively. Adapted from [78]. Reprinted by permission from Springer Nature

was demonstrated by employing a looser focusing geometry and somewhat longer sapphire samples; under these operating conditions, an appreciable redshifted SC signal extended to more than 1600 nm [78, 82], as illustrated in Fig. 5.8a, also see Fig. 4.4 of Sect. 4. A broader, spectrally flat SC was produced using a pump wavelength in the vicinity of the zero GVD point of sapphire (1.31 μm) [83]. Such SC was used to seed the OPCPA system driven by diode-pumped Yb:KGW and Nd:YAG lasers, finally producing carrier envelope phase-stable sub-9 fs pulses with 5.5 TW peak power at a 1 kHz repetition rate [84]. In the range of anomalous GVD of sapphire, a SC spectrum spanning wavelengths from 470 nm to more than 2.5 μm was reported with 15 fs pulses at 2 μm, however, showing a noticeable decrease of the spectral intensity around 1 μm [38].

Another laser host material, YAG, exhibits outstanding mechanical, thermal and optical properties, which make it versatile optical material that is widely used in optoelectronics, laser physics, and nonlinear optics. These properties as combined with a cubic structure of excellent crystalline quality, resulting in the absence of birefringence, in a large nonlinearity and high optical damage threshold makes YAG a very popular nonlinear medium for SC generation. When pumped in the visible spectral range (515 nm), YAG produces a narrower (390–625 nm) and more structured SC spectrum than that obtained in sapphire [77]. However, the advantages of the YAG crystal show up with longer pump wavelengths. With near-infrared pumping, YAG produces quite similar SC spectra as sapphire (420–1600 nm, with 800 nm pumping), which exhibits higher spectral energy density in the wavelength range of 1.0–1.5 μm, as illustrated in Fig. 5.8a, and which is obtained with reduced, sub-μJ pump energies [78] due to the twice larger nonlinear index of refraction of the crystal, see Table 4.1 in Sect. 4. A comparative study of supercontinuum generation by filamentation of 100 fs, 800 nm laser pulses in undoped, Nd-doped and Yb-doped YAG crystals demonstrated that undoped and doped YAG crystals produce almost identical supercontinuum spectra, except for very fine features attributed to the absorption of dopants, i.e., Nd ions in particular, which have multiple absorption lines in the visible and near infrared [85]. Studies of SC generation with wavelength-tunable (in the 1.1–1.6 μm range) pulses from an optical parametric amplifier, demonstrated fairly stable short-wave cutoff at 530 nm and progressive extension of the long-wave (infrared) part of the SC spectrum with increasing the pump wavelengths [78].

With pumping in the mid-infrared, in the range of anomalous GVD (for the pump wavelengths longer than 1.6 μm), YAG shows an advantage over sapphire in producing a more continuous and much flatter SC spectrum over its entire wavelength range. The SC spectra extending from 510 nm to more than 2.5 μm were produced with carrier envelope phase-stable 15 fs pulses at 2 μm [38] and from 450 nm to more than 2.5 μm with 32 fs pulses at 2.15 μm [86], both demonstrating preservation of a stable carrier envelope phase of the broadband radiation. However, in these experiments, the longest detectable wavelength was limited to 2.5 μm. Spectral broadening in YAG to a similar extent was reported with relatively long, 360 fs pulses with a central wavelength of 1937 nm delivered by a newly developed Tm:YAP regenerative amplifier seeded by a Tm:ZBLAN fiber oscillator [87]. The entire extent of the SC spectrum in YAG was measured quite recently with 100 fs, 2.3 μm pump pulses from an optical parametric amplifier, yielding an almost four optical octave-spanning SC with a continuous wavelength coverage from 350 nm to 3.8 μm [40].

With pump pulses of an even longer wavelength, 3.1 μm, delivered by a high repetition rate OPCPA system, a multioctave (from 450 nm to more than 4.5 μm, Fig. 5.9), carrier envelope phase-stable SC was reported, potentially yielding a single optical cycle self-compressed pulses at the output of the crystal, as predicted by the numerical simulations [88]. The authors also performed a polarization analysis of the SC and found that the induced ellipticity is below 0.1 across the entire SC spectrum, showing only a weak dependence on the crystal orientation. This result was confirmed by a systematic investigation of spatio-spectral polarization properties of SC, analyzing both, the axial radiation and conical emission. No depolarization of the supercontinuum, and no spatial dependence of polarization ratios for any wavelength was observed, confirming that polarization is preserved in the process of SC generation in cubic crystals [89].

Fig. 5.9 Supercontinuum spectrum generated by 3.1 μm, 2.6 μJ pulses in 2 mm-thick YAG plate (red curve). The ranges of each spectrometer/detector (Si, InGaAs, and HgCdZnTe) are indicated by color fills. Superimposed is the angle-integrated spectrum from the numerical simulation (blue curve). Adapted from [88]

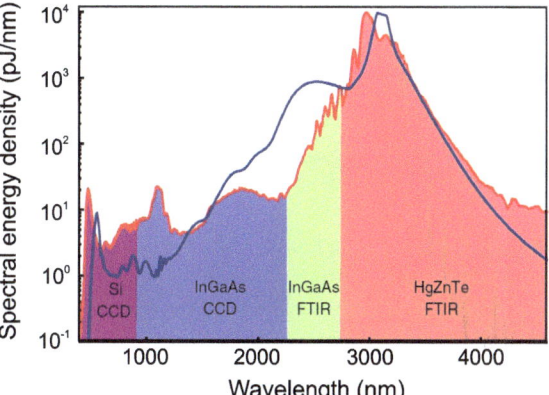

Last but not the least, sapphire and YAG crystals emerge as the only solid-state materials that could be successfully used for SC generation with sub-picosecond and picosecond laser pulses without incurring optical damage. These results are described in detail in Sect. 6.3.

Finally, SC generation was also studied in a number of other laser host materials, such as potassium gadolinium tungstate (KGd(WO$_4$)$_2$, KGW), gadolinium vanadate (GdVO$_4$) and yttrium vanadate (YVO$_4$), being pumped by fundamental harmonics of the Ti:sapphire laser [78]. These crystals exhibit very large nonlinear refractive indexes, which result in very low critical powers for self-focusing, allowing SC generation with sub-100 nJ incident pulse energies using standard 100 fs, 800 nm pump pulses. However, due to relatively small bandgaps and short-wave transmission cut-off wavelengths located in the near ultraviolet, the achieved spectral broadening in these materials was relatively modest. More specifically, SC spectra extending from 500 nm to 1150 nm in KGW, and from 550 to 1150 nm in GdVO$_4$ and YVO$_4$ crystals, were measured, as illustrated in Fig. 5.8b. SC generation with 800–1500 nm tunable pump pulses was also studied in various laser host crystals, such as gadolinium orthosilicate (Gd$_2$SiO$_5$, GSO), gallium gadolinium garnet (Gd$_3$Ga$_5$O$_{12}$, GGG), lithium tantalate (LiTaO$_3$, LTO), and lutetium vanadate (LuVO$_4$, LVO) [90]. The authors measured the anti-Stokes (blueshifted) parts of the SC spectra and recorded fairly constant cutoff wavelengths of 450 nm in GSO and GGG, 550 nm in LTO and 650 nm in LVO crystals. In a later study, spectral characteristics of the Stokes (redshifted) parts of the SC spectra in these crystals were investigated with 800 nm pumping, adding also tetracalcium gadolinium oxoborate (Ca$_4$GdO(BO$_3$)$_3$, GCOB) crystal to the above crystal list [91]. It was demonstrated that the redshifts of the SC spectra increase markedly with increasing energy of the pump pulses. Owing to their relatively large bandgaps ($U_g > 5$ eV), GSO, GCOB, and GGG crystals produced SC spectra extending up to 1500 nm, whereas LVO and LTO crystals possess smaller bandgaps and therefore produced rather modest redshifted broadenings, which were limited to 1300 nm and 1150 nm, respectively.

5.5 Crystals with Second-Order Nonlinearity

Noncentrosymmetric nonlinear crystals which possess second-order nonlinearity and exhibit sufficiently large birefringence to provide phase matching, are indispensable nonlinear media serving for laser wavelength conversion via nonresonant three-wave interactions, such as second harmonic, sum and difference frequency generation, and optical parametric amplification. The joint contribution of quadratic and cubic nonlinearities adds a number of interesting and unique features to the SC generation process in these crystals.

A couple of early works were devoted to study SC generation in noncentrosymmetric crystals neglecting second-order nonlinear effects. To this end, femtosecond filamentation in potassium dihydrogen phosphate (KDP) crystal by launching the pump beam along the optical axis of the crystal was investigated experimentally [92–94] and numerically [95] with a particular emphasis on polarization properties of the generated SC. SC generation was also reported in lithium triborate (LBO) [96], lithium niobate (LN) [97] crystals, and in α-barium borate (α-BBO) crystal, which exhibits birefringence, but vanishing second-order nonlinearity [98]. A series of experiments was carried out under conditions of phase-matched second-order nonlinear processes, reporting SC generation that was accompanied by simultaneous wavelength-tunable second harmonic or sum frequency generation and wavelength-tunable conical emission, as observed in basic nonlinear crystals, such as KDP [99, 100] and β-barium borate (β-BBO) [101, 102], and, more recently, in periodically poled lithium tantalate (PPLT) [103].

A unique aspect of self-action phenomena in birefringent media, which possess both quadratic and cubic nonlinearities, and which was generally neglected in the above studies, is the so-called second-order cascading. The cascading effect arises from the phase mismatched second harmonic generation, which leads to recurrent energy exchange between the fundamental and second harmonic fields, imprinting large nonlinear phase shifts on the interacting waves [104]. The cascading effect mimics the Kerr-like behavior, which arises from the cubic nonlinearity and hence produces a large cascaded nonlinear index of refraction, expressed as:

$$n_2^{\text{casc}} = -\frac{2\omega d_{\text{eff}}^2}{c^2 \epsilon_0 n(\omega)^2 n(2\omega) \Delta k}, \tag{5.1}$$

where d_{eff} is the effective second-order nonlinear coefficient, which sums all relevant components of the quadratic nonlinear susceptibility, ω is the frequency of the fundamental wave, $n(\omega)$ and $n(2\omega)$ are the linear refractive indices of the fundamental and second harmonic waves, respectively, $\Delta k = k(2\omega) - 2k(\omega)$ is the wave vector mismatch (or phase mismatch) parameter, expressed as $k(\omega) = \omega n(\omega)/c$.

The sign and magnitude of the cascaded quadratic nonlinearity can be easily varied by varying the phase mismatch parameter Δk, i.e., by rotating the crystal in the phase matching plane or by changing its temperature. This feature is of particular importance to the nonlinear dynamics of femtosecond pulses [105]. Figure 5.10

Fig. 5.10 **a** Phase mismatch parameter between the fundamental (800 nm) and second harmonic (400 nm) waves in BBO crystal versus the angle θ. **b** Nonlinear refractive indices: intrinsic (n_2^{Kerr}), cascaded (n_2^{casc}), and effective ($n_2^{\mathrm{eff}}=n_2^{\mathrm{Kerr}}+n_2^{\mathrm{casc}}$). The sign and magnitude of the latter indicate relevant self-action regimes

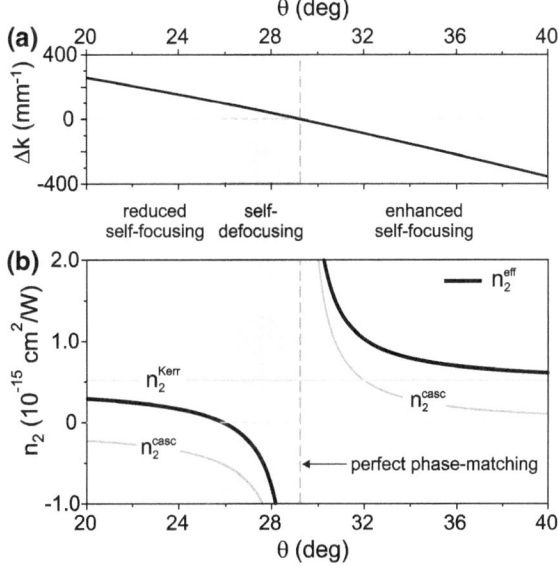

illustrates the phase mismatch parameter, and the calculated cascaded (n_2^{casc}) and effective ($n_2^{\mathrm{eff}} = n_2^{\mathrm{Kerr}} + n_2^{\mathrm{casc}}$) nonlinear indices of refraction of β-BBO crystal as functions of the angle θ, which is the angle between the propagation direction and the optical axis of the crystal.

There are two conceptually different ways how the second-order cascading may be favorably exploited for the generation of broadband spectra in bulk nonlinear crystals. The first approach makes use of the self-defocusing propagation regime, which is achieved for a range of positive phase mismatch parameter values where the cascaded nonlinear index of refraction, n_2^{casc} is negative and its absolute value is greater than the intrinsic n_2^{Kerr}, so $n_2^{\mathrm{eff}} < 0$. Under these operating conditions, spectral broadening is achieved without the onset of beam filamentation and results in pulse self-compression and temporal soliton generation, which in turn emerges from the opposite action of self-phase modulation and material dispersion [106]. Numerical simulations predicted that SC generation regime induced by soliton compression can be achieved in a variety of nonlinear crystals, which possess normal GVD for the pump wavelength [107]. Refer also to Sect. 6.2.2 where soliton compression due to the interplay between the cascaded quadratic and cubic nonlinearities is discussed in more detail. Numerical and experimental results show that the long-wavelength side of the SC radiation could be enriched by the generation of broadband dispersive waves located beyond the zero dispersion wavelength of the crystals: together with the soliton spectrum the dispersive waves contribute to an octave-spanning SC spectrum, which extends from 1.0 to 4.0 μm in LN [108] and from 0.9 to 2.3 μm in β-BBO [109] crystals. The self-defocusing propagation regime by cascaded second-order nonlinearity was demonstrated to be beneficial for the generation of an octave-spanning SC in structured bulk samples. In that regard, a SC spectrum spanning

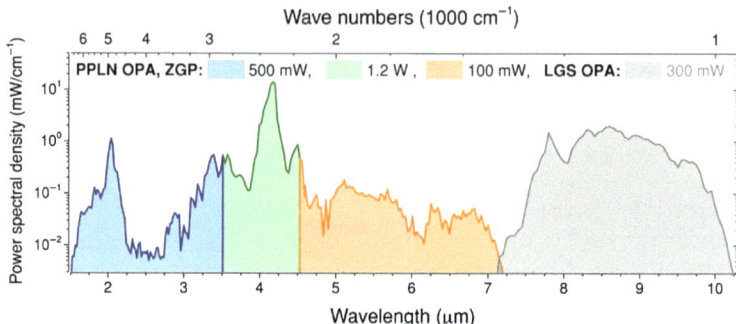

Fig. 5.11 Supercontinuum spectrum generated in a 2 mm-thick ZGP crystal cut at $\theta = 55°$ and pumped by 67 fs pulses at 4.1 µm. The powers of the individual spectral components were determined using a short-pass (blue area) and long-pass (orange area) filters. The gray area shows the spectrum of the LGS optical parametric amplifier. Adapted from [112]

from 1.1 to 2.7 µm was produced in an 11 mm-long periodically poled, Rb-doped potassium titanyl arsenate (KTiOPO$_4$) crystal, using 128 fs, 1.52 µm pump pulses [110]. Alongside SC generation, an almost 7-fold self-compression of the pump pulses down to 18.6 fs was measured.

More recently, this approach was applied to the nonlinear crystals that are transparent in the mid-infrared. To this end, a SC covering the wavelength range from 1.6 to 7.0 µm was experimentally measured in a lithium thioindate (LiInS$_2$) crystal, using pump pulses in the 3–4 µm range [111]. In a similar phase matching configuration, which utilizes the self-defocusing regime, a high average power SC spanning more than two octaves (1.6–7.1 µm) was generated in chalcopyrite (ZnGeP$_2$, ZGP) crystal using incident pulses with a central wavelength of 4.1 µm and duration of 67 fs, delivered by a periodically poled lithium niobate (PPLN) optical parametric amplifier pumped by a Kerr lens mode-locked Yb:YAG thin disc oscillator running at 37.5 MHz repetition rate [112]. If combined with the broadband mid-infrared radiation of LiGaS$_2$ (LGS) optical parametric amplifier, the overall resulting spectrum covers about 2.7 octaves with a reasonable power spectral density (Fig. 5.11).

The second approach makes use of the self-focusing propagation regime, which leads to filamentation of the input beam. Here, the cascaded nonlinearity could be used either to enhance (for negative Δk) or reduce (for positive Δk) the effective nonlinear index of refraction, n_2^{eff}, which is kept positive, see Fig. 5.10. This unique feature opens the possibility to perform the nonlinear interaction in a controlled way. To this end, the interplay between the cascaded quadratic and intrinsic cubic nonlinearities was favorably exploited to achieve filamentation and SC generation with nearly monochromatic, 30 ps pulses from a Nd:YAG laser in a periodically poled LN crystal, also demonstrating control of the SC spectral extent by tuning the crystal temperature [113].

In the femtosecond regime, filamentation of 90 fs, 800 nm laser pulses in β-BBO crystal produced the SC spectrum from 410 nm to 1.1 µm [114]. These experiments

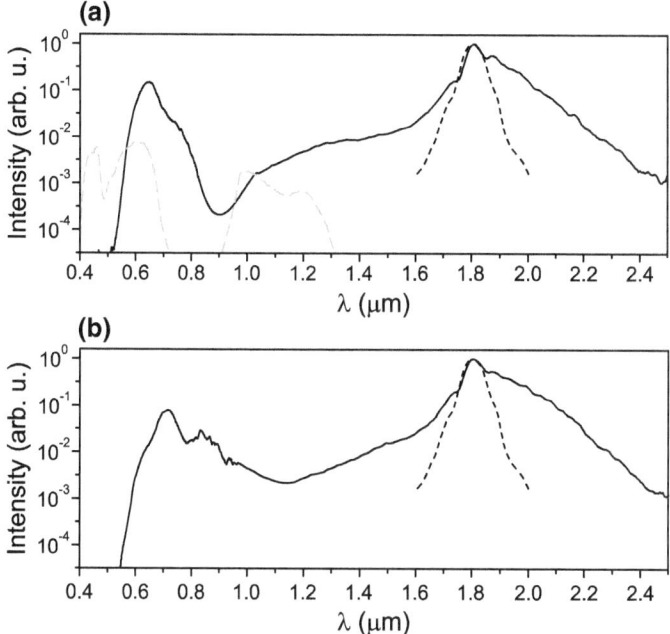

Fig. 5.12 Supercontinuum spectra in a 5-mm-thick β-BBO crystal produced by filamentation of *o*-polarized 100 fs pulses at 1.8 μm for two different crystal orientations yielding **a** enhanced ($\theta = 25.7°$) and **b** reduced ($\theta = 11°$) self-focusing due to second-order cascading. The input pulse energies were 360 nJ and 1.35 μJ, respectively. Spectra of the input pulse and *e*-polarized second harmonic are shown by dotted and dashed curves, respectively

also demonstrated efficient and robust control of the blueshifted portion of the SC spectrum by tuning the angle between the incident laser beam and the optical axis of the crystal, see Fig. 6.13 of Sect. 6.4.

In the region of anomalous GVD of β-BBO crystal, a SC spectrum extending from 520 nm to 2.5 μm was generated with 1.8 μm pump pulses from an optical parametric amplifier, see Fig. 5.12. [115]. In this case, an ultrabroadband SC spectrum was produced by formation of spatiotemporal light bullets, which experience more than 4-fold temporal self-compression from 90 fs down to 20 fs. Moreover, in the θ angle range corresponding to enhanced self-focusing, where n_2^{casc} is positive and larger than n_2^{Kerr}, a markedly reduced filamentation threshold was experimentally detected, allowing SC generation at subcritical powers for self-focusing. On the contrary, in the θ range corresponding to reduced self-focusing, SC generation is achieved with notably elevated pump pulse energies, while the SC spectra in both cases are pretty similar, as illustrated in Fig. 5.12. Interestingly, in the condition of enhanced self-focusing, SC generation is accompanied by efficient generation of orthogonally, *e*-polarized second harmonic, which exhibits a remarkably broad spectral bandwidth and giant spectral shifts from the expected second harmonic wavelength, as shown in Fig. 5.12a. The observed spectral shifts of the second harmonic are attributed to the so-called self-phase matching [116], imposed by the formation of a spatiotemporal light bullet at the fundamental frequency that carries a broadband spectrum. Using soliton terminology, this process is equivalent to the soliton-induced sideband formation at the second harmonic [117].

Similar results regarding the generation of cross-polarized SC components in β-BBO and LN crystals were reported with few optical cycle pump pulses at 2.1 μm [118]. However, due to its smaller bandgap compared to BBO, LN crystal produced a notably narrower SC spectrum, whose blueshifted cutoff was measured just slightly below 800 nm. Spectral broadening mediated by second-order cascading was also reported under different experimental settings. Generation of an intense SC with a spectrum between 1.2 and 3.5 μm and with sub-mJ energy input pulses at 1.5 μm was demonstrated in highly nonlinear organic DAST crystal [119]. A dramatic spectral broadening of sub-two optical cycle pulses with a central wavelength of 790 nm was achieved in β-BBO crystal due to multistep cascaded difference frequency generation producing an ultrabroadband spectrum in the 0.5–2.4 μm range [120].

5.6 Semiconductors

Compared with dielectrics, semiconductor crystals possess much larger cubic non-linearities and extended transparency windows in the mid-infrared, and so emerge as promising nonlinear media for hosting the nonlinear interactions in the near- and mid-infrared spectral ranges, see Table 5.2 that compares relevant optical character-istics of some popular semiconductor materials. SC generation in the mid-infrared spectral range is currently receiving a growing attention due to its application for spectroscopic studies in the so-called molecular fingerprint region as well as for the generation of few optical cycle pulses in this spectral range, opening new experi-mental avenues in strong-field physics. The feasibility of semiconductor materials for SC generation in the mid-infrared spectral range was experimentally demon-strated more than 30 years ago [121]. Here the authors studied spectral broadening

Table 5.2 Relevant linear and nonlinear parameters of some semiconductor materials. U_g is the energy bandgap, transmittance is defined at 10% transmission level of 1 mm thick sample, n_2 is the nonlinear index of refraction, λ_0 is the zero GVD wavelength

Material	U_g eV	Transmittance μm	n_2 $\times 10^{-16}$ cm^2/W	λ_0 μm
ZnS	3.68	0.4–12.5	48	3.6
ZnSe	2.71	0.5–20	60	4.8
GaAs	1.42	0.9–17.3	300	6.0
Si	1.12	1.1–6.5	270	–

of picosecond pulses at 9.3 μm from an amplified CO_2 laser in gallium arsenide (GaAs), zinc selenide (ZnSe), and cadmium sulfide (CdS) crystals. The broadest SC spectrum, continuously covering the 3–14 μm wavelength range was reported in a GaAs crystal. More recently, the SC generation experiment in GaAs crystal was refined using a more modern CO_2 laser system and reported a SC spectrum in the 2–20 μm wavelength range [122].

The real progress regarding SC generation in semiconductor media was achieved quite recently, thanks to the development of femtosecond optical parametric amplifiers and subsequent frequency downconversion techniques, such as difference frequency generation, which made femtosecond mid-infrared pulses routinely available. To this end, a SC spectrum spanning wavelengths from 3.5 to 7 μm was generated in a 10-mm thick GaAs crystal, when pumped by 100 fs pulses with a central wavelength of 5 μm, obtained by difference frequency generation between the signal and idler pulses of near-infrared optical parametric amplifier pumped by a Ti:sapphire laser [123]. A more detailed investigation of SC generation in GaAs crystal in a similar setup was performed using 4.2–6.8 μm tunable pulses and reported spectral broadening in the 4–9 μm range. The authors also reported pulse compression down to sub-two optical cycle widths around 6 μm by performing external post-compression of the spectrally broadened pulses in BaF_2, CaF_2, and MgF_2 crystals, featuring anomalous GVD in this wavelength range [124]. A much broader SC spectrum (3–18 μm) was generated in a GaAs crystal when pumped by femtosecond laser pulses with a central wavelength of 7.9 μm, which falls into the range of anomalous GVD of GaAs [125]. Along with spectral superbroadening, simultaneous pulse self-compression down to almost a single optical cycle was measured.

A series of interesting results were recently reported regarding SC generation in zinc-blende semiconductor crystals. Femtosecond filamentation and spectral broadening were studied in a ZnSe crystal using wavelength-tunable near-infrared pump pulses and so accessing filamentation regimes under different orders of multiphoton absorption [126]. The authors demonstrated that filamentation occurs only in the regimes of three-photon absorption and higher, and observed how spectral broadening increases with increasing the incident wavelength, while no filamentation and spectral broadening was observed with a pump wavelength of 800 nm, which corresponds to the two photon absorption regime in ZnSe. Remarkably broad SC spectra in zinc-blende semiconductor crystals were reported with ultrashort mid-infrared pump pulses. More than three octave-wide supercontinuum, with the spectrum extending from 500 nm to 4.5 μm was generated in a zinc sulfide (ZnS) crystal pumped with 27 fs pulses at 2.1 μm from the OPCPA system [65]. An ultrabroad SC spectrum spanning wavelengths from 500 nm to 11 μm was generated in a 8-mm-long ZnSe crystal using pump pulses of 65 fs duration with a central wavelength of 5 μm, which nearly matched the zero GVD wavelength (4.8 μm) of the crystal [127]. However, such an ultrabroadband SC was produced in the multiple filamentation regime, as verified by the measurements of the near-field intensity distribution of the SC beam.

Owing to the $\bar{4}$3m symmetry, zinc-blende semiconductors possess a nonzero second-order nonlinearity, which could be very favorably exploited to assist spectral

Fig. 5.13 Spectral dynamics versus input pulse energy as produced by filamentation of 100 fs, 2.4 μm pulses in a 5 mm-thick polycrystalline ZnSe sample. Figure on top shows the SC spectrum generated with an input pulse energy of 3 μJ. Adapted from [129]. Reprinted by permission from AIP Publishing

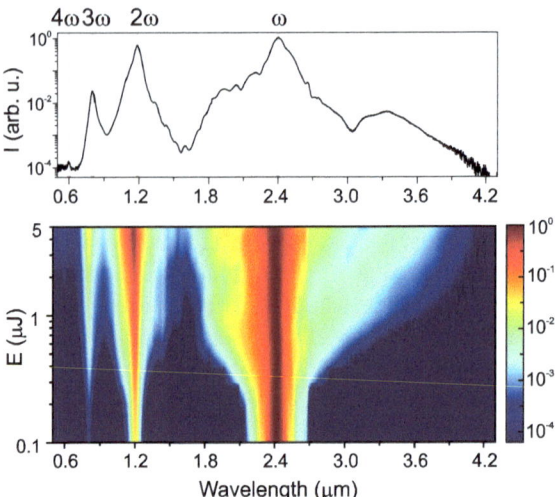

broadening in polycrystalline forms of these materials. The polycrystalline structure provides the so-called random quasi-phase matching, which stems from the disorder of tens-of-microns sized crystallites, greatly extending the limits of frequency conversion that are imposed by the phase mismatch between the interacting waves [128]. Greatly relaxed phase matching conditions enable a broadband frequency conversion within a wide spectral range, therefore polycrystalline forms of zinc-blende semiconductors offer interesting and useful features that accompany filamentation and SC generation processes.

SC generation in a polycrystalline ZnSe sample of 5 mm thickness using 100 fs pump pulses with a carrier wavelength tunable in the 1.5–2.4 μm range was investigated in the regimes of single and multiple filamentation [129]. In particular, setting the pump wavelength at 2.4 μm, efficient generation of the second, third, and fourth harmonics, whose spectral widths replicate the spectral broadening around the carrier wavelength was observed, as illustrated in Fig. 5.13. Prominent broadband spectral peaks at 1.2 μm and 800 nm correspond to the second and third harmonics, respectively, while a weak peak at 600 nm corresponds to the fourth harmonics of the incident wavelength. Note how the spectral broadening around the carrier wavelength is accompanied by the broadening of harmonics spectra, which eventually merge into a broadband SC (shown atop of the Figure) covering the wavelength range from 600 nm to 4.2 μm, that corresponds to 2.8 optical octaves.

The contribution of harmonics generated via random quasi-phase matching to the blueshifted portion of the SC spectrum becomes even more pronounced using pump pulses with longer wavelengths. For instance, harmonic orders up to 6th were produced by filamentation of 85 fs pulses at 3 μm in a 1-mm-thick polycrystalline ZnSe plate [130]. Figure 5.14 presents SC spectra in polycrystalline ZnSe and ZnS

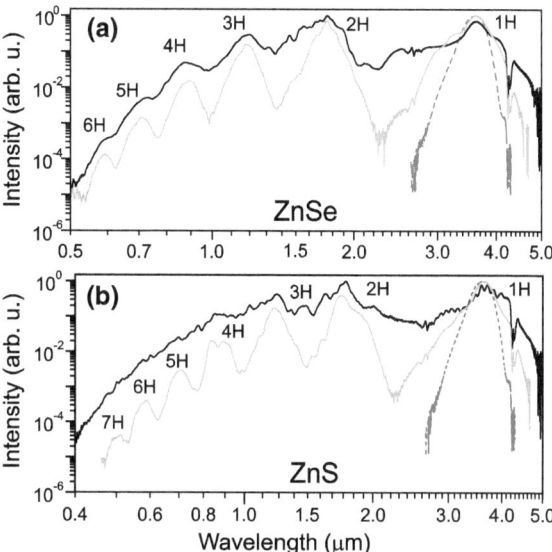

Fig. 5.14 Supercontinuum spectra in polycrystalline **a** ZnSe and **b** ZnS samples, as generated with 60 fs, 3.6 μm pump pulses with an energy of 0.73 μJ. Gray curves depict the SC spectra in 2-mm-thick samples, while solid curves show SC spectra generated in 3-mm-thick ZnSe and 4-mm-thick ZnS samples. The input pulse spectrum is shown by dashed curve. Labels denote the harmonics order. Adapted from [131]. Reprinted by permission from the Optical Society of America

samples of various thickness, produced by filamentation of 60 fs, 0.73 μJ pulses with a carrier wavelength of 3.6 μm [131]. The output spectra produced in shorter (2 mm thick) samples consist of distinct multiple peaks, which correspond to multiple even and odd spectrally broadened harmonics, as labeled on the top. In longer samples, spectral broadening of the individual harmonics and spectral broadening around the carrier wavelength merge, eventually producing an almost uniform ultrabroadband SC spectrum. More specifically, a 3.3 octave-wide SC enhanced by harmonic broadening, spanning the 0.5–5 μm wavelength range, was generated in ZnSe and a 3.6 octave-wide SC spanning the 0.4–5 μm wavelength range was produced in ZnS, whose maximum blue shifts were limited to the short-wavelength transmittance edge of the crystals. These results demonstrate a great potential of polycrystalline zinc-blende semiconductors for SC generation that covers a remarkably broad spectral range from the visible to the mid-infrared.

A recent numerical study of wavelength-scaled filamentation in ZnSe in the range of anomalous GVD of the crystal (for incident wavelengths longer than 4.8 μm, see Table 5.2) uncovered extreme self-steepening and optical shock of mid-infrared and long-wavelength infrared femtosecond pulses, eventually leading to the formation of self-compressed light bullets with sub-cycle temporal widths [132]. Such extreme pulse compression was demonstrated to produce SC spectra spanning multiple octaves and extending well beyond 15 μm on the long-wavelength side.

5.7 Other Nonlinear Media

A couple of other nonlinear materials were tested for SC generation. Only a moderate spectral broadening of Ti:sapphire laser pulses at 800 nm was reported in dielectric crystals possessing relatively narrow bandgaps, such as diamond ($U_g = 5.47$ eV). Here, the measured SC spectrum showed a clear signature of stimulated Raman scattering, while the measured spectral blueshift extended just slightly below 600 nm [133]. Under similar pumping conditions, somewhat broader SC spectra were reported in calcite ($CaCO_3$), which has a slightly larger bandgap ($U_g = 5.9$ eV) [134]. An interesting experimental study on SC generation with femtosecond pulses at 1.24 μm from a Cr:forsterite laser was performed in various aggregate states of CO_2: high-pressure gas, liquid, and supercritical fluid [135]. The authors reported SC generation in the 0.4–2.2 μm range, and demonstrated that the SC spectral shape is strongly dependent on pressure.

A comprehensive numerical study on SC generation in a variety of solid-state nonlinear materials was performed with pump wavelengths set slightly above their zero GVD points [136]. Numerical simulations predicted that filamentation of femtosecond mid-infrared pulses in alkali metal halides, which possess extremely broad mid-infrared transmittance and whose zero GVD points are located deeply in the mid-infrared, are capable of producing multioctave SC spectra with remarkable red shifts. For instance, the simulated SC spectra in sodium chloride (NaCl) and potassium iodide (KI) using pump pulses with a carrier wavelength of 5 μm, cover the wavelength ranges of 0.7–7.6 μm and 0.66–22 μm, respectively, demonstrating a great, and yet experimentally unexplored potential of these materials for SC generation in the mid- and far-infrared spectral range.

To this end, a more recent experimental study has identified mixed thallium halides: thallium bromoiodide (KRS-5) and thallium chlorobromide (KRS-6) as attractive nonlinear materials, which have a great potential for SC generation in the mid-infrared spectral range [137]. KRS-5 and KRS-6 are crystals with cubic structure and are well-known as infrared window materials with relatively narrow bandgaps (2.5 eV and 3.25 eV, respectively) and transparency ranges of 0.58–42 μm and 0.42–27 μm, respectively, which extend into the far infrared. Almost 2 octave-spanning SC spectra, covering the ∼1.5–5.5 μm wavelength range, were produced by filamentation of 60 fs pulses at 3.1 μm in KRS-5 and KRS-6 samples of 6 mm thickness, as illustrated in Fig. 5.15. Owing to large nonlinear refractive indexes of refraction: 10.5×10^{-15} cm^2/W for KRS-5 and 5.4×10^{-15} cm^2/W for KRS-6, as evaluated from measurements of the nonlinear transmission, filamentation and SC generation in these materials was achieved with remarkably low input pulse energies: just several hundreds of nanojoules were required to produce a single filament, while pulses with an energy of ∼1 μJ readily yielded beam break up into multiple filaments. In light of these results, generation of even broader SC spectra could be

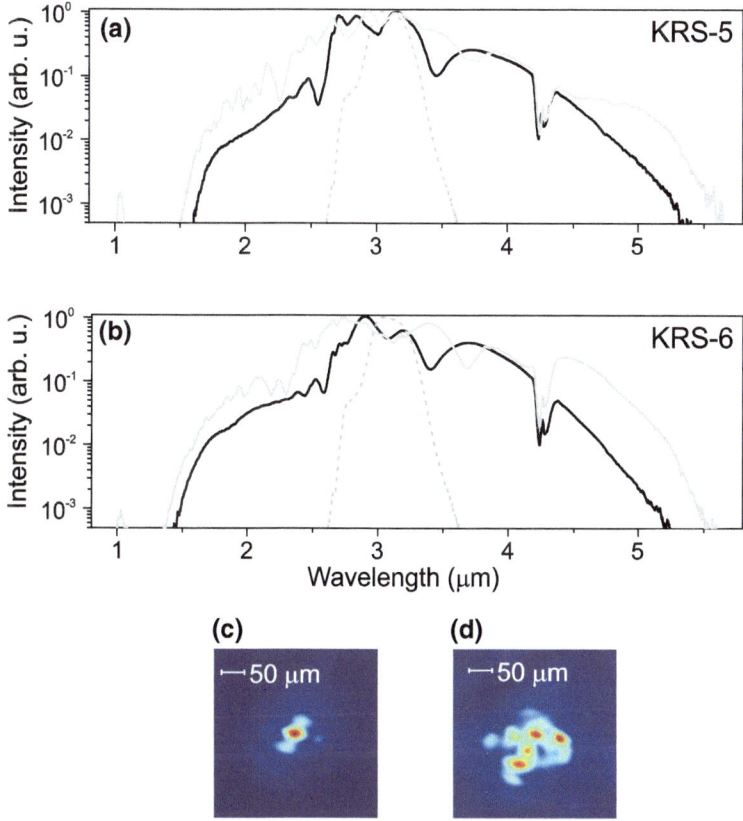

Fig. 5.15 Supercontinuum spectra in **a** KRS-5 and **b** KRS-6 crystals of 6 mm thickness, generated by self-focusing of 60 fs pulses with a carrier wavelength of 3.1 μm in the regimes of single (black curves) and multiple (gray curves) filamentation. The input pulse spectrum is shown by the dashed curve. The narrow peaks at 1.03 μm correspond to the third harmonics of the pump. **c** and **d** show typical intensity distributions of single and multiple filaments, respectively, measured at the output of KRS-5 crystal. Adapted from [137]. Reprinted by permission from Elsevier

expected using femtosecond pump pulses with longer wavelengths, suggesting KRS-5 and KRS-6 crystals as very attractive alternatives to soft glasses and semiconductor crystals for SC generation in the mid-infrared spectral range.

References

1. Smith, W.L., Liu, P., Bloembergen, N.: Superbroadening in H_2O and D_2O by self-focused picosecond pulses from a YAlG: Nd laser. Phys. Rev. A **15**, 2396–2403 (1977)
2. Golub, I.: Optical characteristics of supercontinuum generation. Opt. Lett. **15**, 305–307 (1990)

3. He, G.S., Xu, G.C., Cui, Y., Prasad, P.N.: Difference of spectral superbroadening behavior in Kerr-type and non-Kerr-type liquids pumped with ultrashort laser pulses. Appl. Opt. **32**, 4507–4512 (1993)
4. Brodeur, A., Ilkov, F.A., Chin, S.L.: Beam filamentation and the white light continuum divergence. Opt. Commun. **129**, 193–198 (1996)
5. Wittmann, M., Penzkofer, A.: Spectral supebroadening of femtosecond laser pulses. Opt. Commun. **126**, 308–317 (1996)
6. Nagura, C., Suda, A., Kawano, H., Obara, M., Midorikawa, K.: Generation and characterization of ultrafast white-light continuum in condensed media. Appl. Opt. **41**, 3735–3742 (2002)
7. Jarnac, A., Tamošauskas, G., Majus, D., Houard, A., Mysyrowicz, A., Couairon, A., Dubietis, A.: Whole life cycle of femtosecond ultraviolet filaments in water. Phys. Rev. A **89**, 033809 (2014)
8. Dubietis, A., Tamošauskas, G., Diomin, I., Varanavičius, A.: Self-guided propagation of femtosecond light pulses in water. Opt. Lett. **28**, 1269–1271 (2003)
9. Liu, W., Kosareva, O., Golubtsov, I.S., Iwasaki, A., Becker, A., Kandidov, V.P., Chin, S.L.: Random deflection of the white light beam during self-focusing and filamentation of a femtosecond laser pulse in water. Appl. Phys. B **75**, 595–599 (2002)
10. Liu, W., Kosareva, O., Golubtsov, I.S., Iwasaki, A., Becker, A., Kandidov, V.P., Chin, S.L.: Femtosecond laser pulse filamentation versus optical breakdown in H_2O. Appl. Phys. B **76**, 215–229 (2003)
11. Kandidov, V.P., Kosareva, O.G., Golubtsov, I.S., Liu, W., Becker, A., Aközbek, N., Bowden, C.M., Chin, S.L.: Self-transformation of a powerful femtosecond laser pulse into a white-light laser pulse in bulk optical media (or supercontinuum generation). Appl. Phys. B **77**, 149–165 (2003)
12. Cook, K., Kar, A.K., Lamb, R.A.: White-light supercontinuum interference of self-focused filaments in water. Appl. Phys. Lett. **83**, 3861–3863 (2003)
13. Liu, J., Schroeder, H., Chin, S.L., Li, R., Xu, Z.: Nonlinear propagation of fs laser pulses in liquids and evolution of supercontinuum generation. Opt. Express **13**, 10248–10259 (2005)
14. Tcypkin, A.N., Putilin, S.E., Melnik, M.V., Makarov, E.A., Bespalov, V.G., Kozlov, S.A.: Generation of high-intensity spectral supercontinuum of more than two octaves in a water jet. Appl. Opt. **55**, 8390–8394 (2016)
15. Potemkin, F.V., Mareev, E.I., Smetanina, E.O.: Influence of wavefront curvature on supercontinuum energy during filamentation of femtosecond laser pulses in water. Phys. Rev. A **97**, 033801 (2018)
16. Vasa, P., Dharmadhikari, J.A., Dharmadhikari, A.K., Sharma, R., Singh, M., Mathur, D.: Supercontinuum generation in water by intense, femtosecond laser pulses under anomalous chromatic dispersion. Phys. Rev. A **89**, 043834 (2014)
17. Dharmadhikari, J.A., Steinmeyer, G., Gopakumar, G., Mathur, D., Dharmadhikari, A.K.: Femtosecond supercontinuum generation in water in the vicinity of absorption bands. Opt. Lett. **41**, 3475–3478 (2016)
18. Tzortzakis, S., Papazoglou, D.G., Zergioti, I.: Long-range filamentary propagation of subpicosecond ultraviolet laser pulses in fused silica. Opt. Lett. **31**, 796–798 (2006)
19. Nguyen, N.T., Saliminia, A., Liu, W., Chin, S.L., Vallée, R.: Optical breakdown versus filamentation in fused silica by use of femtosecond infrared laser pulses. Opt. Lett. **28**, 1591–1593 (2003)
20. Ashcom, J.B., Gattass, R.R., Schaffer, C.B., Mazur, E.: Numerical aperture dependence of damage and supercontinuum generation from femtosecond laser pulses in bulk fused silica. J. Opt. Soc. Am. B **23**, 2317–2322 (2006)
21. Fang, X.-J., Kobayashi, T.: Evolution of a super-broadened spectrum in a filament generated by an ultrashort intense laser pulse in fused silica. Appl. Phys. B **77**, 167–170 (2003)
22. Dachraoui, H., Oberer, C., Michelswirth, M., Heinzmann, U.: Direct time-domain observation of laser pulse filaments in transparent media. Phys. Rev. A **82**, 043820 (2010)
23. Zhang, L., Xi, T., Hao, Z., Lin, J.: Supercontinuum accumulation along a single femtosecond filament in fused silica. J. Phys. D **49**, 115201 (2016)

24. Zafar, S., Li, D., Hao, Z., Lin, J.: Influences of astigmatic focusing geometry on femtosecond filamentation and supercontinuum generation in fused silica. Optik **130**, 765–769 (2017)
25. Dharmadhikari, A.K., Rajgara, F.A., Mathur, D.: Systematic study of highly efficient white-light generation in transparent materials using intense femtosecond pulses. Appl. Phys. B **80**, 61–66 (2005)
26. Dharmadhikari, A.K., Rajgara, F.A., Mathur, D.: Depolarization of white light generated by ultrashort laser pulses in optical media. Opt. Lett. **31**, 2184–2186 (2006)
27. Yang, J., Mu, G.: Multi-dimensional observation of white-light filaments generated by femtosecond laser pulses in condensed medium. Opt. Express **15**, 4943–4952 (2007)
28. Jiang, J., Zhong, Y., Zheng, Y., Zeng, Z., Ge, X., Li, R.: Broadening of white-light continuum by filamentation in BK7 glass at its zero-dispersion point. Phys. Lett. A **379**, 1929–1933 (2015)
29. Dubietis, A., Tamošauskas, G., Šuminas, R., Jukna, V., Couairon, A.: Ultrafast supercontinuum generation in bulk condensed media (Review). Lith. J. Phys. **57**, 113–157 (2017)
30. Saliminia, A., Chin, S.L., Vallée, R.: Ultra-broad and coherent white light generation in silica glass by focused femtosecond pulses at 1.5 μm. Opt. Express **13**, 5731–5738 (2005)
31. Naudeau, M.L., Law, R.J., Luk, T.S., Nelson, T.R., Cameron, S.M.: Observation of nonlinear optical phenomena in air and fused silica using a 100 GW, 1.54 μm source. Opt. Express **14**, 6194–6200 (2006)
32. Faccio, D., Averchi, A., Couairon, A., Dubietis, A., Piskarskas, R., Matijošius, A., Bragheri, F., Porras, M.A., Piskarskas, A., Di Trapani, P.: Competition between phase-matching and stationarity in Kerr-driven optical pulse filamentation. Phys. Rev. E **74**, 047603 (2006)
33. Porras, M.A., Dubietis, A., Matijošius, A., Piskarskas, R., Bragheri, F., Averchi, A., Di Trapani, P.: Characterization of conical emission of light filaments in media with anomalous dispersion. J. Opt. Soc. Am. B **24**, 581–584 (2007)
34. Durand, M., Lim, K., Jukna, V., McKee, E., Baudelet, M., Houard, A., Richardson, M., Mysyrowicz, A., Couairon, A.: Blueshifted continuum peaks from filamentation in the anomalous dispersion regime. Phys. Rev. A **87**, 043820 (2013)
35. Smetanina, E.O., Kompanets, V.O., Chekalin, S.V., Dormidonov, A.E., Kandidov, V.P.: Anti-Stokes wing of femtosecond laser filament supercontinuum in fused silica. Opt. Lett. **38**, 16–18 (2013)
36. Chekalin, S.V., Kompanets, V.O., Dokukina, A.E., Dormidonov, A.E., Smetanina, E.O., Kandidov, V.P.: Visible supercontinuum radiation of light bullets in the femtosecond filamentation of IR pulses in fused silica. Quantum Electron. **45**, 401–407 (2015)
37. Gražulevičiūtė, I., Šuminas, R., Tamošauskas, G., Couairon, A., Dubietis, A.: Carrier-envelope phase-stable spatiotemporal light bullets. Opt. Lett. **40**, 3719–3722 (2015)
38. Darginavičius, J., Majus, D., Jukna, V., Garejev, N., Valiulis, G., Couairon, A., Dubietis, A.: Ultrabroadband supercontinuum and third-harmonic generation in bulk solids with two optical-cycle carrier-envelope phase-stable pulses at 2 μm. Opt. Express **21**, 25210–25220 (2013)
39. Dharmadhikari, J.A., Deshpande, R.A., Nath, A., Dota, K., Mathur, D., Dharmadhikari, A.K.: Effect of group velocity dispersion on supercontinuum generation and filamentation in transparent solids. Appl. Phys. B **117**, 471–479 (2014)
40. Garejev, N., Tamošauskas, G., Dubietis, A.: Comparative study of multioctave supercontinuum generation in fused silica, YAG, and LiF in the range of anomalous group velocity dispersion. J. Opt. Soc. Am. B **34**, 88–94 (2017)
41. Yang, Y., Liao, M., Li, X., Bi, W., Ohishi, Y., Cheng, T., Fang, Y., Zhao, G., Gao, W.: Filamentation and supercontinuum generation in lanthanum glass. J. Appl. Phys. **121**, 023107 (2017)
42. Liao, M., Gao, W., Cheng, T., Xue, X., Duan, Z., Deng, D., Kawashima, H., Suzuki, T., Ohishi, Y.: Five-octave-spanning supercontinuum generation in fluoride glass. Appl. Phys. Express **6**, 032503 (2013)
43. Liao, M., Gao, W., Cheng, T., Duan, Z., Xue, X., Kawashima, H., Suzuki, T., Ohishi, Y.: Ultrabroad supercontinuum generation through filamentation in tellurite glass. Laser Phys. Lett. **10**, 036002 (2013)

44. Béjot, P., Billard, F., Peureux, C., Diard, T., Picot-Clémente, J., Strutynski, C., Mathey, P., Mouawad, O., Faucher, O., Nagasaka, K., Ohishi, Y., Smektala, F.: Filamentation-induced spectral broadening and pulse shortening of infrared pulses in Tellurite glass. Opt. Commun. **380**, 245–249 (2016)

45. Yu, Y., Gai, X., Wang, T., Ma, P., Wang, R., Yang, Z., Choi, D.-Y., Madden, S., Luther-Davies, B.: Mid-infrared supercontinuum generation in chalcogenides. Opt. Mater. Express **3**, 1075–1086 (2013)

46. Mouawad, O., Béjot, P., Billard, F., Mathey, P., Kibler, B., Désévédavy, F., Gadret, G., Jules, J.-C., Faucher, O., Smektala, F.: Mid-infrared filamentation-induced supercontinuum in As-S and an As-free Ge-S counterpart chalcogenide glasses. Appl. Phys. B **121**, 433–438 (2015)

47. Stingel, A.M., Vanselous, H., Petersen, P.B.: Covering the vibrational spectrum with microjoule mid-infrared supercontinuum pulses in nonlinear optical applications. J. Opt. Soc. Am. B **34**, 1163–1168 (2017)

48. Mouawad, O., Béjot, P., Mathey, P., Froidevaux, P., Lemière, A., Billard, F., Kibler, B., Désévédavy, F., Gadret, G., Jules, J.-C., Faucher, O., Smektala, F.: Expanding up to far-infrared filamentation-induced supercontinuum spanning in chalcogenide glasses. Appl. Phys. B **124**, 182 (2018)

49. Riedle, E., Bradler, M., Wenninger, M., Sailer, C.F., Pugliesi, I.: Electronic transient spectroscopy from the deep UV to the NIR: unambiguous disentanglement of complex processes. Faraday Discuss. **163**, 139–158 (2013)

50. Ziolek, M., Naskrecki, R., Karolczak, J.: Some temporal and spectral properties of femtosecond supercontinuum important in pump-probe spectroscopy. Opt. Commun. **241**, 221–229 (2004)

51. Kartazaev, V., Alfano, R.R.: Polarization properties of SC generated in CaF_2. Opt. Commun. **281**, 463–468 (2008)

52. Johnson, P.J.M., Prokhorenko, V.I., Miller, R.J.D.: Stable UV to IR supercontinuum generation in calcium fluoride with conserved circular polarization states. Opt. Express **17**, 21488–21496 (2009)

53. Tzankov, P., Buchvarov, I., Fiebig, T.: Broadband optical parametric amplification in the near UV-VIS. Opt. Commun. **203**, 107–113 (2002)

54. Wang, J., Zhang, Y., Shen, H., Jiang, Y., Wang, Z.: Spectral stability of supercontinuum generation in condensed mediums. Opt. Eng. **56**, 076107 (2017)

55. Buchvarov, I., Trifonov, A., Fiebig, T.: Toward an understanding of white-light generation in cubic media-polarization properties across the entire spectral range. Opt. Lett. **32**, 1539–1541 (2007)

56. Zeller, J., Jaspara, J., Rudolph, W., Sheik-Bahae, M.: Spectro-temporal characterization of a femtosecond white-light continuum by transient grating diffraction. Opt. Commun. **185**, 133–137 (2000)

57. Megerle, U., Pugliesi, I., Schriever, C., Sailer, C.F., Riedle, E.: Sub-50 fs broadband absorption spectroscopy with tunable excitation: putting the analysis of ultrafast molecular dynamics on solid ground. Appl. Phys. B **96**, 215–231 (2009)

58. Krebs, N., Pugliesi, I., Hauer, J., Riedle, E.: Two-dimensional Fourier transform spectroscopy in the ultraviolet with sub-20 fs pump pulses and 250–720 nm supercontinuum probe. New. J. Phys. **15**, 085016 (2013)

59. Huber, R., Satzger, H., Zinth, W., Wachtveitl, J.: Noncollinear optical parametric amplifiers with output parameters improved by the applications of a white light continuum generated in CaF_2. Opt. Commun. **194**, 443–448 (2001)

60. Dharmadhikari, A.K., Rajgara, F.A., Reddy, N.C.S., Sandhu, A.S., Mathur, D.: Highly efficient white light generation from barium fluoride. Opt. Express **12**, 695–700 (2004)

61. Dharmadhikari, A.K., Alti, K., Dharmadhikari, J.A., Mathur, D.: Control of the onset of filamentation in condensed media. Phys. Rev. A **76**, 033811 (2007)

62. Dharmadhikari, A.K., Rajgara, F.A., Mathur, D.: Plasma effects and the modulation of white light spectra in the propagation of ultrashort, high-power laser pulses in barium fluoride. Appl. Phys. B **82**, 575–583 (2006)

63. Dharmadhikari, A.K., Dharmadhikari, J.A., Mathur, D.: Visualization of focusing-refocusing cycles during filamentation in BaF_2. Appl. Phys. B **94**, 259–263 (2009)
64. Harth, A., Schultze, M., Lang, T., Binhammer, T., Rausch, S., Morgner, U.: Two-color pumped OPCPA system emitting spectra spanning 1.5 octaves from VIS to NIR. Opt. Express **20**, 3076–3081 (2012)
65. Liang, H., Krogen, P., Grynko, R., Novak, O., Chang, C.-L., Stein, G.J., Weerawarne, D., Shim, B., Kärtner, F.X., Hong, K.-H.: Three-octave-spanning supercontinuum generation and sub-two-cycle self-compression of mid-infrared filaments in dielectrics. Opt. Lett. **40**, 1069–1072 (2015)
66. Dormidonov, A.E., Kompanets, V.O., Chekalin, S.V., Kandidov, V.P.: Giantically blue-shifted visible light in femtosecond mid-IR filament in fluorides. Opt. Express **23**, 29202–29210 (2015)
67. Chekalin, S.V., Kompanets, V.O., Dormidonov, A.E., Zaloznaya, E.D., Kandidov, V.P.: Supercontinuum spectrum upon filamentation of laser pulses under conditions of strong and weak anomalous group velocity dispersion in transparent dielectrics. Quantum Electron. **47**, 252–258 (2017)
68. Marcinkevičiūtė, A., Garejev, N., Šuminas, R., Tamošauskas, G., Dubietis, A.: A compact, self-compression-based sub-3 optical cycle source in the 3–4 μm spectral range. J. Opt. **19**, 105505 (2017)
69. Kohl-Landgraf, J., Nimsch, J.-E., Wachtveitl, J.: LiF, an underestimated supercontinuum source in femtosecond transient absorption spectroscopy. Opt. Express **21**, 17060–17065 (2013)
70. Dormidonov, A.E., Kompanets, V.O., Chekalin, S.V., Kandidov, V.P.: Dispersion of the anti-stokes band in the spectrum of a light bullet of a femtosecond filament. JETP Lett. **104**, 175–179 (2016)
71. Chekalin, S.V., Kompanets, V.O., Dormidonov, A.E., Kandidov, V.P.: Influence of induced colour centres on the frequency-angular spectrum of a light bullet of mid-IR radiation in lithium fluoride. Quantum Electron. **47**, 259–265 (2017)
72. Yang, Y., Bi, W., Li, X., Liao, M., Gao, W., Ohishi, Y., Fang, Y., Li, Y.: Ultrabroadband supercontinuum generation through filamentation in a lead fluoride crystal. J. Opt. Soc. Am. B **36**, A1–A7 (2019)
73. Reed, M.K., Steiner-Shepard, M.K., Negus, D.K.: Widely tunable femtosecond optical parametric amplifier at 250 kHz with a Ti:sapphire regenerative amplifier. Opt. Lett. **19**, 1855–1857 (1994)
74. Yakovlev, V.V., Kohler, B., Wilson, K.R.: Broadly tunable 30-fs pulses produced by optical parametric amplification. Opt. Lett. **19**, 2000–2002 (1994)
75. Reed, M.K., Steiner-Shepard, M.K., Armas, M.S., Negus, D.K.: Microjoule-energy ultrafast optical parametric amplifiers. J. Opt. Soc. Am. B **12**, 2229–2236 (1995)
76. Manzoni, C., Cerullo, G.: Design criteria for ultrafast optical parametric amplifiers. J. Opt. **18**, 103501 (2016)
77. Bradler, M., Riedle, E.: Sub-20 fs μJ-energy pulses tunable down to the near-UV from a 1 MHz Yb-fiber laser system. Opt. Lett. **39**, 2588–2591 (2014)
78. Bradler, M., Baum, P., Riedle, E.: Femtosecond continuum generation in bulk laser host materials with sub-μJ pump pulses. Appl. Phys. B **97**, 561–574 (2009)
79. Majus, D., Jukna, V., Pileckis, E., Valiulis, G., Dubietis, A.: Rogue-wave-like statistics in ultrafast white-light continuum generation in sapphire. Opt. Express **19**, 16317–16323 (2011)
80. Majus, D., Dubietis, A.: Statistical properties of ultrafast supercontinuum generated by femtosecond Gaussian and Bessel beams: a comparative study. J. Opt. Soc. Am. B **30**, 994–999 (2013)
81. Imran, T., Figueira, G.: Intensity-phase characterization of white-light continuum generated in sapphire by 280 fs laser pulses at 1053 nm. J. Opt. **14**, 035201 (2012)
82. Jukna, V., Galinis, J., Tamošauskas, G., Majus, D., Dubietis, A.: Infrared extension of femtosecond supercontinuum generated by filamentation in solid-state media. Appl. Phys. B **116**, 477–483 (2014)
83. Budriūnas, R., Stanislauskas, T., Varanavičius, A.: Passively CEP-stabilized frontend for few cycle terawatt OPCPA system. J. Opt. **17**, 094008 (2015)

84. Budriūnas, R., Stanislauskas, T., Adamonis, J., Aleknavičius, A., Veitas, G., Gadonas, D., Balickas, S., Michailovas, A., Varanavičius, A.: 53 W average power CEP-stabilized OPCPA system delivering 5.5 TW few cycle pulses at 1 kHz repetition rate. Opt. Express **25**, 5797–5806 (2017)

85. Kudarauskas, D., Tamošauskas, G., Vengris, M., Dubietis, A.: Filament-induced luminescence and supercontinuum generation in undoped, Yb-doped and Nd-doped YAG crystals. Appl. Phys. Lett. **112**, 041103 (2018)

86. Fattahi, H., Wang, H., Alismail, A., Arisholm, G., Pervak, V., Azzeer, A.M., Krausz, F.: Near-PHz-bandwidth, phase-stable continua generated from a Yb:YAG thin-disk amplifier. Opt. Express **24**, 24337–24346 (2016)

87. Rezvani, S.A., Suzuki, M., Malevich, P., Livache, C., De Montgolfier, J.V., Nomura, Y., Tsurumachi, N., Baltuška, A., Fuji, T.: Millijoule femtosecond pulses at 1937 nm from a diode-pumped ring cavity Tm:YAP regenerative amplifier. Opt. Express **26**, 29460–29470 (2018)

88. Silva, F., Austin, D.R., Thai, A., Baudisch, M., Hemmer, M., Faccio, D., Couairon, A., Biegert, J.: Multi-octave supercontinuum generation from mid-infrared filamentation in a bulk crystal. Nat. Commun. **3**, 807 (2012)

89. Choudhuri, A., Chatterjee, G., Zheng, J., Hartl, I., Ruehl, A., Miller, R.J.D.: A spatio-spectral polarization analysis of 1 μm-pumped bulk supercontinuum in a cubic crystal (YAG). Appl. Phys. B. **124**, 103 (2018)

90. Ryba-Romanowski, W., Macalik, B., Strzęp, A., Lisiecki, R., Solarz, P., Kowalski, R.M.: Spectral transformation of infrared ultrashort pulses in laser crystals. Opt. Mater. **36**, 1745–1748 (2014)

91. Macalik, B., Kowalski, R.M., Ryba-Romanowski, W.: Spectral features of the Stokes part of supercontinuum generated by femtosecond light pulses in selected oxide crystals: a comparative study. Opt. Mater. **78**, 396–401 (2018)

92. Kumar, R.S.S., Deepak, K.L.N., Rao, D.N.: Control of the polarization properties of the supercontinuum generation in a noncentrosymmetric crystal. Opt. Lett. **33**, 1198–1200 (2008)

93. Kumar, R.S.S., Deepak, K.L.N., Rao, D.N.: Depolarization properties of the femtosecond supercontinuum generated in condensed media. Phys. Rev. A **78**, 043818 (2008)

94. Yu, J., Jiang, H., Yang, H., Gong, Q.: Depolarization of white light generated by femtosecond laser pulse in KDP crystals. J. Opt. Soc. Am. B **28**, 1566–1570 (2011)

95. Rolle, J., Bergé, L., Duchateau, G., Skupin, S.: Filamentation of ultrashort laser pulses in silica glass and KDP crystals: a comparative study. Phys. Rev. A **90**, 023834 (2014)

96. Faccio, D., Di Trapani, P., Minardi, S., Bramati, A., Bragheri, F., Liberale, C., Degiorgio, V., Dubietis, A., Matijosius, A.: Far-field spectral characterization of conical emission and filamentation in Kerr media. J. Opt. Soc. Am. B **22**, 862–869 (2005)

97. Wang, Y., Ni, H., Zhan, W., Yuan, J., Wang, R.: Supercontinuum and THz generation from Ni implanted $LiNbO_3$ under 800 nm laser excitation. Opt. Commun. **291**, 334–336 (2013)

98. Vasa, P., Dota, K., Singh, M., Kushavah, D., Singh, B.P., Mathur, D.: Power- and polarization-dependent supercontinuum generation in α-BaB_2O_4 crystals by intense, near-infrared, femtosecond laser pulses. Phys. Rev. A **91**, 053837 (2015)

99. Srinivas, N.K.M.N., Harsha, S.S., Rao, D.N.: Femtosecond supercontinuum generation in a quadratic nonlinear medium (KDP). Opt. Express **13**, 3224–3229 (2005)

100. Kumar, R.S.S., Harsha, S.S., Rao, D.N.: Broadband supercontinuum generation in a single potassium di-hydrogen phosphate (KDP) crystal achieved in tandem with sum frequency generation. Appl. Phys. B **86**, 615–621 (2007)

101. Wang, L., Fan, Y.X., Zhu, H., Yan, Z.D., Zeng, H., Wang, H.-T., Zhu, S.N., Wang, Z.L.: Broadband colored-crescent generation in a single β-barium-borate crystal by intense femtosecond pulses. Phys. Rev. A **84**, 063831 (2011)

102. Ali, S.A., Bisht, P.B., Nautiyal, A., Shukla, V., Bindra, K.S., Oak, S.M.: Conical emission in β-barium borate under femtosecond pumping with phase matching angles away from second harmonic generation. J. Opt. Soc. Am. B **27**, 1751–1756 (2010)

103. Zhao, L., Zeng, X., Tong, L., Gao, Y., Liu, J., Ren, Y., Zhao, Y., Li, J.: Femtosecond supercontinuum generation and Čerenkov conical emission in periodically poled $LiTaO_3$. Optik **156**, 333–337 (2018)

104. Stegeman, G.I., Hagan, D.J., Torner, L.: $\chi^{(2)}$ cascading phenomena and their applications to all-optical signal processing, mode-locking, pulse compression and solitons. Opt. Quantum Electron. **28**, 1691–1740 (1996)
105. Wise, F.W., Moses, J.: Self-focusing and self-defocusing of femtosecond pulses with cascaded quadratic nonlinearities. Top. Appl. Phys. **114**, 481–506 (2009)
106. Zhou, B.B., Chong, A., Wise, F.W., Bache, M.: Ultrafast and octave-spanning optical nonlinearities from strongly phase-mismatched quadratic interactions. Phys. Rev. Lett. **109**, 043902 (2012)
107. Bache, M., Guo, H., Zhou, B.: Generating mid-IR octave-spanning supercontinua and few-cycle pulses with solitons in phase-mismatched quadratic nonlinear crystals. Opt. Mater. Express **3**, 1647–1657 (2013)
108. Zhou, B., Guo, H., Bache, M.: Energetic mid-IR femtosecond pulse generation by self-defocusing soliton-induced dispersive waves in a bulk quadratic nonlinear crystal. Opt. Express. **23**, 6924–6936 (2015)
109. Zhou, B., Bache, M.: Dispersive waves induced by self-defocusing temporal solitons in a beta-barium-borate crystal. Opt. Lett. **40**, 4257–4260 (2015)
110. Viotti, A.-L., Lindberg, R., Zukauskas, A., Budriunas, R., Kucinskas, D., Stanislauskas, T., Laurell, F., Pasiskevicius, V.: Supercontinuum generation and soliton self-compression in $\chi^{(2)}$-structured KTiOPO$_4$. Optica **5**, 711–717 (2018)
111. Zhou, B., Bache, M.: Multiple-octave spanning mid-IR supercontinuum generation in bulk quadratic nonlinear crystals. APL Photon. **1**, 050802 (2016)
112. Seidel, M., Xiao, X., Hussain, S.A., Arisholm, G., Hartung, A., Zawilski, K.T., Schunemann, P.G., Habel, F., Trubetskov, M., Pervak, V., Pronin, O., Krausz, F.: Multi-watt, multi-octave, mid-infrared femtosecond source. Sci. Adv. **4**, eaaq1526 (2018)
113. Krupa, K., Labruyère, A., Tonello, A., Shalaby, B.M., Couderc, V., Baronio, F., Aceves, A.B.: Polychromatic filament in quadratic media: spatial and spectral shaping of light in crystals. Optica **2**, 1058–1064 (2015)
114. Šuminas, R., Tamošauskas, G., Jukna, V., Couairon, A., Dubietis, A.: Second-order cascading-assisted filamentation and controllable supercontinuum generation in birefringent crystals. Opt. Express **25**, 6746–6756 (2017)
115. Šuminas, R., Tamošauskas, G., Valiulis, G., Dubietis, A.: Spatiotemporal light bullets and supercontinuum generation in β-BBO crystal with competing quadratic and cubic nonlinearities. Opt. Lett. **41**, 2097–2100 (2016)
116. Valiulis, G., Jukna, V., Jedrkiewicz, O., Clerici, M., Rubino, E., Di Trapani, P.: Propagation dynamics and X-pulse formation in phase-mismatched second-harmonic generation. Phys. Rev. A **83**, 043834 (2011)
117. Zhou, B., Guo, H., Bache, M.: Soliton-induced nonlocal resonances observed through high-intensity tunable spectrally compressed second-harmonic peaks. Phys. Rev. A **90**, 013823 (2014)
118. Wang, H., Alismail, A., Barbiero, G., Wendl, M., Fattahi, H.: Cross-polarized, multi-octave supercontinuum generation. Opt. Lett. **42**, 2595–2598 (2017)
119. Vicario, C., Monoszlai, B., Arisholm, G., Hauri, C.P.: Generation of 1.5-octave intense infrared pulses by nonlinear interactions in DAST crystal. J. Opt. **17**, 094005 (2015)
120. Kessel, A., Trushin, S.A., Karpowicz, N., Skrobol, C., Klingebiel, S., Wandt, C., Karsch, S.: Generation of multi-octave spanning high-energy pulses by cascaded nonlinear processes in BBO. Opt. Express **24**, 5628–5637 (2016)
121. Corkum, P.B., Ho, P.P., Alfano, R.R., Manassah, J.T.: Generation of infrared supercontinuum covering 3–14 μm in dielectrics and semiconductors. Opt. Lett. **10**, 624–626 (1985)
122. Pigeon, J.J., Tochitsky, S.Ya., Gong, C., Joshi, C.: Supercontinuum generation from 2 to 20 μm in GaAs pumped by picosecond CO$_2$ laser pulses. Opt. Lett. **39**, 3246–3249 (2014)
123. Ashihara, S., Kawahara, Y.: Spectral broadening of mid-infrared femtosecond pulses in GaAs. Opt. Lett. **34**, 3839–3841 (2009)
124. Lanin, A.A., Voronin, A.A., Stepanov, E.A., Fedotov, A.B., Zheltikov, A.M.: Frequency-tunable sub-two-cycle 60-MW-peak-power free-space waveforms in the mid-infrared. Opt. Lett. **39**, 6430–6433 (2014)

125. Lanin, A.A., Voronin, A.A., Stepanov, E.A., Fedotov, A.B., Zheltikov, A.M.: Multioctave, 3–18 μm sub-two-cycle supercontinua from self-compressing, self-focusing soliton transients in a solid. Opt. Lett. **40**, 974–977 (2015)

126. Durand, M., Houard, A., Lim, K., Durécu, A., Vasseur, O., Richardson, M.: Study of filamentation threshold in zinc selenide. Opt. Express **22**, 5852–5858 (2014)

127. Mouawad, O., Béjot, P., Billard, F., Mathey, P., Kibler, B., Désévédavy, F., Gadret, G., Jules, J.-C., Faucher, O., Smektala, F.: Filament-induced visible-to-mid-IR supercontinuum in a ZnSe crystal: towards multi-octave supercontinuum absorption spectroscopy. Opt. Mater. **60**, 355–358 (2016)

128. Baudrier-Raybaut, M., Haïdar, R., Kupecek, Ph., Lemasson, Ph., Rosencher, E.: Random quasi-phase-matching in bulk polycrystalline isotropic nonlinear materials. Nature **432**, 374–376 (2004)

129. Šuminas, R., Tamošauskas, G., Valiulis, G., Jukna, V., Couairon, A., Dubietis, A.: Multi-octave spanning nonlinear interactions induced by femtosecond filamentation in polycrystalline ZnSe. Appl. Phys. Lett. **110**, 241106 (2017)

130. Archipovaite, G.M., Petit, S., Delagnes, J.-C., Cormier, E.: 100 kHz Yb-fiber laser pumped 3 μm optical parametric amplifier for probing solid-state systems in the strong field regime. Opt. Lett. **42**, 891–894 (2017)

131. Šuminas, R., Marcinkevičiūtė, A., Tamošauskas, G., Dubietis, A.: Even and odd harmonics-enhanced supercontinuum generation in zinc-blende semiconductors. J. Opt. Soc. Am. B **36**, A22–A27 (2019)

132. Grynko, R.I., Nagar, G.C., Shim, B.: Wavelength-scaled laser filamentation in solids and plasma-assisted subcycle light-bullet generation in the long-wavelength infrared. Phys. Rev. A **98**, 023844 (2018)

133. Kardaś, T.M., Ratajska-Gadomska, B., Gadomski, W., Lapini, A., Righini, R.: The role of stimulated Raman scattering in supercontinuum generation in bulk diamond. Opt. Express **21**, 24201–24209 (2013)

134. Kartazaev, V., Alfano, R.R.: Supercontinuum generated in calcite with chirped femtosecond pulses. Opt. Lett. **32**, 3293–3295 (2007)

135. Mareev, E., Bagratashvili, V., Minaev, N., Potemkin, F., Gordienko, V.: Generation of an adjustable multi-octave supercontinuum under near-IR filamentation in gaseous, supercritical, and liquid carbon dioxide. Opt. Lett. **41**, 5760–5763 (2016)

136. Frolov, S.A., Trunov, V.I., Leshchenko, V.E., Pestryakov, E.V.: Multi-octave supercontinuum generation with IR radiation filamentation in transparent solid-state media. Appl. Phys. B **122**, 124 (2016)

137. Marcinkevičiūtė, A., Tamošauskas, G., Dubietis, A.: Supercontinuum generation in mixed thallous halides KRS-5 and KRS-6. Opt. Mater. **78**, 339–344 (2018)

Chapter 6
New Developments

Emerging applications in many fields of modern ultrafast science require high-power broadband radiation and few optical cycle pulses that are generated by all-solid-state technology and by relatively simple means. This chapter gives an overview of the currently evolving research topics in SC generation that pursue power and energy scaling, and control of SC radiation. A special emphasis is given to various extra-cavity pulse compression and self-compression techniques which rely on spectral broadening and SC generation in bulk solid-state media, and which currently receive a great deal of revived interest. The physical picture of picosecond filamentation and the experimental achievements reporting a series of exciting results on SC generation with picosecond laser pulses in solids without incurring optical damage of the material are provided in detail. At the end, experimental results on SC generation with non-Gaussian (Bessel, Airy, and vortex) ultrashort-pulsed pump beams are summarized. Finally, the techniques enabling broadband frequency up- and down-conversion of the SC spectrum are presented.

6.1 Power and Energy Scaling

The energy, power, and spectral density of the SC radiation in the single filamentation regime is limited by the intensity clamping effect, and these practically important parameters could not be scaled by simple means. The straightforward increase of the input pulse energy leads to beam break-up into multiple filaments, and the resulting SC spectrum is a superposition of the SC spectra produced by individual filaments. Although the formation of multiple filaments creates an illusion of power and energy scaling of the SC radiation, it is achieved at the cost of unwanted degradation of spatial and temporal coherence, see Sect. 4.5, therefore restricting the area of applications of such broadband light.

Recently, a novel concept for the generation of high-power, high-energy SC in a solid-state medium was proposed and demonstrated by using a series of distributed thin plates instead of a single thick piece of bulk nonlinear medium [1]. The idea

© The Author(s), under exclusive license to Springer Nature Switzerland AG 2019
A. Dubietis and A. Couairon, *Ultrafast Supercontinuum Generation
in Transparent Solid-State Media*, SpringerBriefs in Physics,
https://doi.org/10.1007/978-3-030-14995-6_6

Fig. 6.1 a Output spectrum as a function of the fused silica plate number. Each plate has a thickness of 0.1 mm. The far-field images of the beam after spectral broadening in **b** a sequence of distributed four 0.1-mm-thick fused silica plates, **c** a single piece of fused silica of 0.5 mm thickness. Adapted from [1]. Reprinted by permission from the Optical Society of America

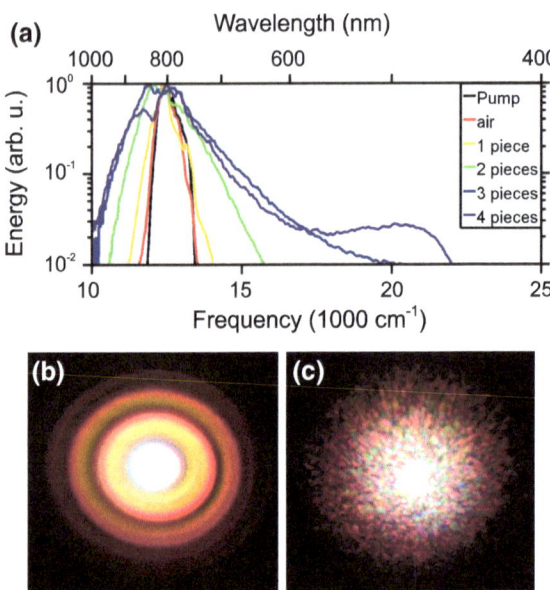

of this concept is that individual thin plates of the nonlinear medium are distributed in space so that the pulse exits each plate with a broadened spectrum, but before the onset of small-scale self-focusing and break-up of the high-power beam into multiple filaments. The small-scale irregularities of the beam are then washed out by diffraction during free space propagation, before the beam enters the next plate. In that way, after passing a sequence of plates, broadband radiation with high energy and with a uniform spatial profile is produced at the output. Figure 6.1a shows experimentally measured spectra of the incident 25 fs, 140 μJ laser pulses at 800 nm in a sequence of 100 − μm-thick fused silica plates as functions of the plate number, finally yielding an octave-spanning SC spectrum, which covers the wavelength range from 450 to 980 nm after the fourth and final plate. At the output of the multi-plate arrangement, the broadband radiation has a smooth far-field intensity distribution, as shown in Fig. 6.1b, whose central spot contains 54% of the incident energy. This configuration yields an almost 100-fold increase in spectral density over existing bulk solid-state sources that employ a single filament for SC generation. For a comparison, Fig. 6.1c shows the far-field image of the SC beam generated by self-focusing in a single piece of fused silica of 0.5 mm thickness, showing the occurrence of a large number of hotspots indicating beam break-up into multiple filaments.

The operation of the multi-plate arrangement was further studied by performing numerical simulations and experiments, which demonstrate the scaling capability of the input peak power up to as much as two thousand times the critical power for self-focusing in a solid-state medium, keeping the overall energy transmission above 50% [2]. It was also found that the principal limitation in power scaling comes from nonlinear effects in the atmospheric air, when the peak power of the input pulse

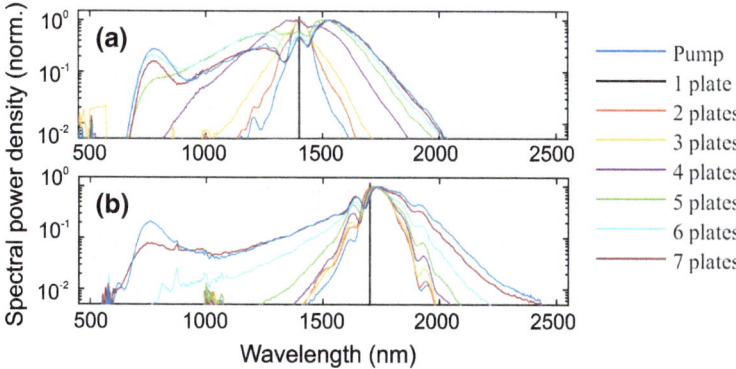

Fig. 6.2 Supercontinuum spectra generated in an arrangement of multiple alternating fused silica and sapphire plates of 200 μm thickness each, with pump wavelengths of **a** 1400 nm, **b** 1700 nm, which fall into the ranges of near-zero and anomalous GVD of the nonlinear media, respectively. Adapted from [4]. Reprinted by permission from Springer Nature

exceeds the filamentation threshold for air before the pulse enters the first plate. In the course of further development of the multi-plate technique, energy scaling of the SC radiation to sub-millijoule (0.8 mJ) energy level was demonstrated with 30 fs laser pulses at 790 nm in a sequence of seven fused silica plates with almost 90% throughput efficiency [3].

The advantages of the multi-plate concept were also demonstrated by the generation of high-power broadband SC with wavelength-tunable pulses from an optical parametric amplifier. More specifically, SC generation was performed in a sequence of seven alternating UV-grade fused silica and sapphire plates of 200 μm thickness using 60 fs input pulses with central wavelengths in the 1.2−1.75 μm range, so accessing the regimes of near-zero to anomalous GVD of fused silica and sapphire and with the pump pulse energies up to 0.5 mJ [4]. Figure 6.2 shows the SC spectra as functions of the plate number, recorded with pump pulse wavelengths of 1.4 and 1.7 μm, which fall into the ranges of near-zero and anomalous GVD of the non-linear media, respectively. An experimental comparison between the SC generated in the multiple plate arrangement and continuous nonlinear medium showed that the energy scaling is achieved without the degradation of the essential performance characteristics, such as spectral width, pulse-to-pulse energy stability, and stability of the carrier-envelope phase of the output pulses.

Remarkably, the cascaded spectral broadening in a distributed multi-plate arrangement was demonstrated to yield the SC pulse with a regular chirp, which is compressible to a nearly transform-limited pulse using an external pulse compressor [1, 3]. Applications of the multi-plate concept for various schemes of extracavity pulse compression are discussed in more detail in Sect. 6.2.1.

6.2 Extracavity Pulse Compression

Extracavity pulse compression is based on increasing the spectral bandwidth via self-phase modulation (SPM) in a nonlinear medium and subsequent removal of the frequency modulation by using an appropriate dispersive delay line. This constitutes a simple and robust method for obtaining few optical cycle pulses whose spectral widths extend well beyond the gain bandwidth supported by the ultrashort pulse lasers and optical parametric amplifiers. Historically, SPM-induced spectral broadening in single-mode fibers in the wavelength ranges of normal and anomalous GVD, in combination with either an external pulse compressor or making use of the soliton compression effect, respectively, has been serving as an efficient technique for production of the shortest pulses in the optical range for many years [5–7]. However, due to its small core area, a single-mode fiber is limited by material damage and supports the propagation of very low (few nJ) energy pulses, and therefore many other variants of pulse compression schemes relying on various nonlinear media have been proposed since.

The advent of hollow-core fibers filled with noble gases at high-pressure boosted remarkable progress in the development of extracavity pulse compression techniques [8]. The hollow-fiber technique enabled the achievement of spectral broadening of high-energy (of the order of several milijoules) laser pulses preserving a uniform spatial profile. It took advantage of the chirped-mirror technology that offers dispersion engineering in the compression stage. Recent developments of this compression technique are dedicated to further optimization of the compression setups aiming at the reduction of unwanted nonlinear effects by creating a pressure gradient across the beam path [9] or by employing photonic crystal fibers with a large hollow core filled with a noble gas as a nonlinear optical medium [10]. On the other hand, self-compression through filamentation of femtosecond pulses in the atmosphere was demonstrated in the ranges of normal [11] and anomalous [12] GVD of air and, due to its simplicity, may be considered as an attractive alternative to fiber-based compression setups.

Pulse compression based on SPM-induced spectral broadening in a bulk solid-state medium offers the advantages of a wide variety of suitable nonlinear materials, technical simplicity, low cost, and easy implementation to virtually any existing ultrashort pulse laser system, but requires proper managing of nonlinear effects in the spatial domain [13, 14]. Moreover, several methods of pulse compression that are experimentally very attractive have been elaborated, where no additional dispersive elements to compensate SPM-induced frequency modulation are required, as both spectral broadening and temporal compression are performed within a single piece of nonlinear material. This type of self-compression mechanism is often termed as "soliton compression", in analogy with temporal soliton generation in optical fibers. The first method exploits the interplay between SPM via cascaded self-defocusing second-order nonlinearity in non-centrosymmetric crystals and normal GVD of the medium. The second method makes use of the SPM-induced spectral broadening in bulk solid-state media with cubic nonlinearity featuring anomalous GVD, which can be performed with or without beam filamentation.

In the meantime, we are witnessing a revival of pulse compression techniques, which pursue the generation of few optical cycle pulses at various parts of the optical spectrum. These techniques are based on all-solid-state technology and inspired by the development of novel ultrafast lasers and ultrashort-pulsed optical parametric amplifiers. Recent experimental achievements suggest that extracavity pulse compression employing spectral broadening in a bulk nonlinear medium could be easily scaled in wavelength, and readily applied for a range of incident pulse widths varying from sub-100 fs durations to a few picoseconds.

6.2.1   Pulse Compression Exploiting SPM in Normally Dispersive Media

The vast majority of the ultrashort-pulsed lasers emit in the near infrared, where dielectric bulk materials exhibit normal GVD, and therefore pulse compression is performed in a "classical" way: the SPM-induced pulse chirp is removed by using an appropriate dispersive delay line (pulse compressor), in which the optical paths are organized so as to delay the fastest (red-shifted) and to advance the slowest (blue-shifted) spectral components. This pulse compression technique attracts an increased practical interest, in particular due to its applicability to novel Yb-based lasers, which produce either relatively long (few hundreds of fs) femtosecond or sub-picosecond pulses with very high average power. To this end, a sixfold spectral broadening by self-phase modulation in a single 15-mm-thick quartz crystal and subsequent chirped-mirror compression factors of about 6 with 60% throughput efficiency were demonstrated with 250 fs pulses at 1.03 μm with 1.3 μJ energy and an average power of 50 W at a repetition rate of 38 MHz from a mode-locked thin-disk Yb:YAG oscillator [15]. Numerical simulations with input pulses of the same parameters launched into the multi-plate setup predicted an even larger spectral broadening and compressibility of the pulses down to 15 fs. In particular, these numerical simulations revealed the useful role of the air gaps between the plates, which lead to homogenizing of the spectrum over the beam profile and significantly reduce the nonlinear losses. A combination of fiber and bulk compression techniques using the same laser oscillator pulses yielded waveform-stabilized 7.7 fs (2.2 optical cycles) pulses with an energy of 0.15 μJ and an average power of 6 W [16]. More specifically, here, a two-step spectral broadening was performed in an 8-cm-long large-mode-area photonic crystal fiber and crystalline quartz plate of 5 mm thickness, each followed by a chirped-mirror compressor.

A comprehensive study of the bulk compression in the ultraviolet spectral range was performed in [17]. The third harmonic output of a Ti:sapphire amplifier (257 nm, 120 fs, 1.5 μJ) was spectrally broadened in a 1-mm-thick CaF_2 plate and compressed down to 50 fs using a simple prism compressor, yielding a compression factor of 2.5 and energy transmission of 70%. The spectral broadening by a factor of ~1.5 in CaF_2 plates of 2 and 1 mm thickness was demonstrated with sub-30 fs frequency up-converted UV pulses from the visible noncollinear parametric amplifier at 273 nm and 313 nm, respectively. Compression of these pulses to nearly their Fourier limit was performed by implementing an acousto-optical pulse shaper, yielding the output pulses with durations of 20 fs and 15 fs at 273 nm and 313 nm, respectively. However, only 10% overall energy transmission was measured due to high-energy losses introduced by the shaper device.

Thanks to a wide choice of suitable nonlinear materials, bulk compression could be efficiently performed in the mid-infrared spectral range as well. In that regard, spectra spanning one octave were generated via SPM-induced spectral broadening in a 7-mm-thick GaAs crystal using incident pulses with $80-190$ fs duration and carrier wavelengths tunable in the $4.2-6.8$ μm range, obtained through difference frequency generation by mixing the signal and idler waves from an optical parametric amplifier [18]. Spectrally broadened pulses were afterward compressed by the propagation in CaF_2, BaF_2, and MgF_2 plates of millimetric thickness, which provide anomalous group velocity dispersion in this wavelength range. A suitable combination of different materials allowed precise tailoring of the dispersion profile over the entire spectral bandwidth. In particular, more than fourfold compression of 120 fs, 5.9 μm input pulses was demonstrated, yielding sub-two optical cycle (29 fs) pulses with a few μJ energy.

A combination of bulk compression and multi-plate concepts holds a great potential for energy scaling of the compressed pulses. In that regard, an almost 10-fold increase of the spectral bandwidth of sub-picosecond (0.85 ps) pulses from an Yb:YAG-Innoslab amplifier was reported in a multi-pass cell geometry, while maintaining almost diffraction-limited beam quality [19]. A complete optical path comprised 38 passes through a fused silica substrate with a total propagation length of $\sim$1 m in the nonlinear medium and $\sim$20 m in free space. The spectrally broadened pulses were compressed down to 170 fs by a dispersive mirror compressor, yielding the output pulses with 375 W average power at repetition rate of 10 MHz. A relatively simple and elegant demonstration of the multi-plate concept for pulse compression was reported in an experiment where input 0.8 mJ, 30 fs laser pulses at 790 nm were spectrally broadened in a sequence of seven identical fused silica plates of 100 μm thickness each, and compressed using chirped mirrors down to 5.4 fs (two optical cycles) with almost 90% throughput efficiency [3]. 280 fs pulses with a moderately high energy (400 μJ) and average power (20 W), with a central wavelength of 1025 nm from an Yb:KGW laser, were spectrally broadened in a multi-plate medium consisting of 6 mm of fused silica in total and compressed down to 50 fs using chirped mirrors [20]. Addition of yet another cascade of a multi-plate medium that also consisted of 6 mm of fused silica in total, enabled further compression of the pulses down to 18 fs.

A record spectral broadening by a factor of 22 was demonstrated in a Herriott-type multi-pass cell, which uses a single, large-aperture nonlinear medium (fused silica) allowing tens of passes through it with a compact footprint. Here, the output pulses from a high-power thin-disk Yb:YAG oscillator were compressed from 220 fs to 18 fs, while maintaining an efficiency of over 60% at an average input power of 100 W,

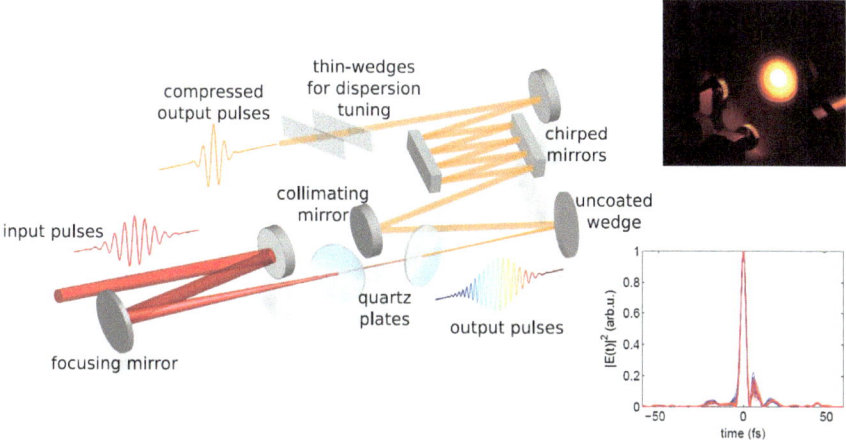

Fig. 6.3 Experimental setup for pulse compression down to a single-cycle limit. Two thin quartz plates (of 50 μm thickness each) placed at Brewster angle around the focal plane of the loosely focused beam are used to extend the pulse spectrum to more than a full octave. Compression is achieved with chirped mirrors and thin fused silica wedges. The insets show a photograph of the resulting beam with octave-spanning spectrum and the temporal profile of the compressed pulse. Adapted from [22]

and an excellent output beam quality with $M^2 = 1.2$ [21]. The broadest measured spectrum supports a Fourier transform limit of 10 fs.

More recently, pulse compression down to a single optical cycle limit was demonstrated in a very compact, simple, and cost-effective setup, which is depicted in Fig. 6.3. The spectrum of 7 fs pulses with a central wavelength of 800 nm from a high average power (10 W), high repetition rate (100 kHz) noncollinear optical parametric chirped pulse amplifier is broadened to beyond one octave by exploiting self-phase modulation in a pair of thin quartz plates and air. Pulse compression was performed utilizing multiple bounces off chirped mirrors in a combination with material dispersion of the air and thin fused silica wedges, yielding compressed pulses with 3.7 fs duration (1.5 optical cycle) and a peak power of 26 GW, with excellent wavefront characteristics and stable carrier-envelope phase [22].

Eventually, numerical simulations predict that multiple thin plates of the solid-state medium separated by appropriate spatial filters could be employed to broaden the spectrum of ultrahigh peak power pulses, yielding spectral bandwidths that are compressible to few-cycle pulse widths, with undistorted output beam profiles that are focusable to relativistic and ultrarelativistic intensities [23]. Very recently, experiments performed with the HERCULES laser at the Center for Ultrafast Optical Science (CUOS) at the University of Michigan Ann Arbor, and the LASERIX facility at the Université Paris-Sud in Orsay, France, reported on spectral broadening in fused silica wafers of 0.5 mm thickness and compression of intense 55 fs pulses with 296 mJ energy to 31 fs [24].

6.2.2 Soliton Compression Due to Second-Order Cascading

Historically, the first experimental demonstration of soliton compression in a bulk medium dates back to 1995 [25], where the authors exploited the interplay between self-phase modulation via second-order cascading due to phase-mismatched second harmonic generation in a crystal with second-order nonlinearity (β-BBO), and normal GVD. A unique aspect of the second-order cascading is the possibility to obtain large cascaded Kerr-like nonlinearity that can be adjusted in sign and magnitude by varying the phase mismatch parameter (see Sect. 5.5 for more details). In particular, this provides access to the self-defocusing propagation regime, where the total (or the so-called effective) nonlinear refractive index is negative. As a result, the SPM effect stemming from negative effective nonlinear index of refraction imprints a frequency modulation that is compressible by the normal GVD of the material. Since the self-focusing effect is readily eliminated in the spatial domain, this approach has no fundamental limitation for the input pulse energy scaling, while the practical limitations come just from the availability of suitable large-aperture nonlinear crystals with second-order nonlinearity. This regime of soliton compression was widely exploited for obtaining few optical cycle pulses in the near infrared, in the region of normal GVD of popular nonlinear crystals, such as BBO [26–28] and LN [29]. More recently, an octave-spanning SC generation and a remarkable, almost sevenfold soliton compression of 128 fs pulses at 1.52 µm down to 18.6 fs was demonstrated in an 11-mm-long periodically poled, Rb-doped potassium titanyl arsenate (PPKTP) crystal [30]. The power scaling of soliton compression was demonstrated using a commercial Kerr-lens mode-locked thin-disk Yb:YAG oscillator which delivers 190 fs pulses with 90 W average power. Using a BBO crystal, soliton compression of the oscillator pulses down to 30 fs with 70 W average power was reported [31].

Interestingly, spectral broadening and SC generation in the soliton compression regime is accompanied by the generation of a broadband dispersive wave, which lies in the long-wavelength side of the spectrum, where the material GVD is anomalous [32, 33]. Measurements show that this dispersive wave is chirped; therefore, using an external pulse compressor, the dispersive wave alone could be compressed to the transform limit, i.e., down to a few optical cycles, as suggested by its broad spectral width. A comprehensive theoretical and numerical study predicted that such soliton compression regime could be attained in the near- and mid-infrared spectral ranges using a variety of nonlinear crystals with second-order nonlinearity [34].

Favorable conditions for soliton compression in crystals with the second-order nonlinearity may be also satisfied in the spectral region where GVD is anomalous. In contrast to the soliton compression mechanism that involves second-order cascading to achieve the self-defocusing propagation regime in the range of normal GVD, this particular case makes use of the second-order cascading to produce a large effective self-focusing nonlinearity that facilitates self-compression in the range of anomalous GVD. As in the case of isotropic nonlinear media (see the next section), beam filamentation effects are avoided by choosing an appropriate input beam diameter,

Fig. 6.4 Experimentally measured **a** temporal and **b** spectral dynamics of 68 fs, 73 μJ input pulses at 2.1 μm in a 5.5-mm-thick BBO crystal versus the crystal rotation angle. The FWHM diameter of the input beam is 1.03 mm, intensity 84 GW/cm^2. Vertical line indicates the perfect phase-matching angle for second harmonic generation ($\theta_{pm} = 21.6°$). Adapted from [35]. Reprinted by permission from the Optical Society of America

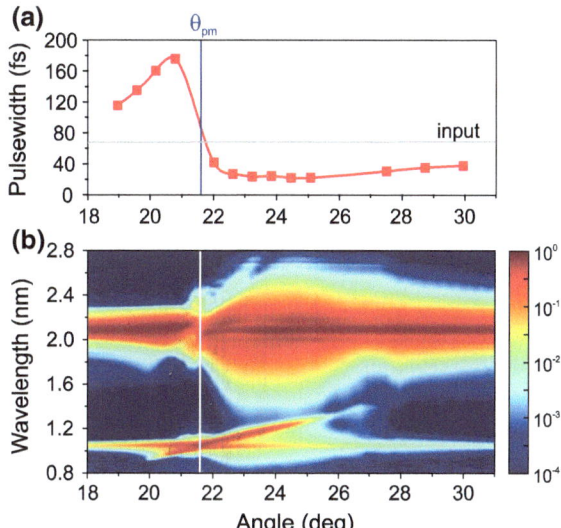

which guarantees an essentially one-dimensional (temporal) dynamics leading to soliton compression without the onset of beam filamentation and associated nonlinear losses.

Figure 6.4 shows the experimentally measured temporal and spectral dynamics of 68 fs, 73 μJ input pulses in a 5.5-mm-thick BBO crystal versus the crystal rotation angle [35]. The carrier wavelength (2.1 μm) of the input pulse was in the region of anomalous GVD of the crystal (the zero GVD is at 1.49 μm), and the self-compression was observed when the crystal was rotated out of the phase matching for second harmonic generation ($\theta_{pm} = 21.6°$) so as to induce a large positive effective nonlinear index of refraction. A more than threefold filamentation-free self-compression from 68 to 22 fs was measured at 24.5° with an energy throughput of 86.3%, where the energy losses originated just from Fresnel reflections from an uncoated crystal sample and phase-mismatched second harmonic generation. A threefold energy scaling using the same setup was demonstrated as well, yielding a self-compressed pulse with a 4.5 GW peak power.

6.2.3 Self-compression Through Filamentation

Filamentation of femtosecond laser pulses in isotropic solid-state materials featuring anomalous GVD provides a natural condition for pulse self-compression. In contrast to filamentation in normally dispersive media, in the present case, the red-shifted and blue-shifted frequencies that are generated on the ascending and descending pulse fronts, respectively, are pushed back to the peak of the pulse by the anomalous GVD of the medium. As a consequence, a simultaneous compression in the temporal and spatial domains occurs, giving rise to the formation of self-compressed light bullets, which do not spread in time and space over long propagation distances [37]. To this end, self-compression of near- and mid-infrared femtosecond pulses down to a few optical cycles was experimentally demonstrated in the $1.8-3.9$ μm range in a variety of transparent dielectric media featuring anomalous GVD, such as fused silica [36–39], sapphire[40], YAG [41, 42], CaF_2, BaF_2 [43], LiF [44], and BBO [45]. However, despite its simplicity, self-compression through filamentation in anomalously dispersive media has several drawbacks, which may appear undesirable for a range of practical applications. First, filamentation does not allow energy scaling of the self-compressed pulses and incurs significant energy losses due to multiphoton and plasma absorption. Second, an input Gaussian-shaped wave packet undergoes dramatic reshaping in the spatial and temporal domains, so the self-compressed pulse carries just a relatively small fraction of the incident energy [40].

6.2.4   Soliton Compression in Isotropic Nonlinear Media with Anomalous GVD

A more practical realization of the anomalous GVD-induced self-compression regime in isotropic nonlinear media relies upon using a shorter nonlinear medium and a larger input beam, i.e., decoupling the temporal and spatial dynamics, which then distinctly exhibit different length scales. As a result, the self-compressed pulse is extracted before the catastrophic self-focusing and beam filamentation set in, minimizing or completely avoiding nonlinear energy losses, hence ensuring high-energy throughput and maintaining a homogenous beam profile of the self-compressed pulse. The possibility to self-compress the pulses by spectral broadening in media with anomalous GVD was foreseen in the seminal paper of Rolland and Corkum [13], but its practical realizations had to wait until suitable femtosecond laser sources operating in the near- and mid-infrared were developed.

Numerical simulations predicted that filamentation-free self-compression in isotropic media could yield pulse widths approaching a single optical cycle as long as conditions for a one-dimensional propagation dynamics of the input pulse are maintained [46–48]. By choosing appropriately the nonlinear material, the input beam size and intensity, soliton compression could be relatively easily scaled in wavelength and energy [49–51].

The feasibility of scaling of the soliton compression in solid-state media to multimillijoule energies and terawatt peak powers was experimentally demonstrated with a hybrid OPA/OPCPA system operating at 3.9 μm. Here self-compression from 94 fs down to 30 fs (sub-three optical cycles) was achieved as a result of a favorable interplay between a strong anomalous GVD and optical nonlinearity of a 2-mm-thick YAG crystal located before the focal plane of a focusing lens [52], as illustrated in Fig. 6.5. Energy losses during the self-compression process were measured to be as low as 4%, and the self-compressed pulse maintained a homogenous spatial profile

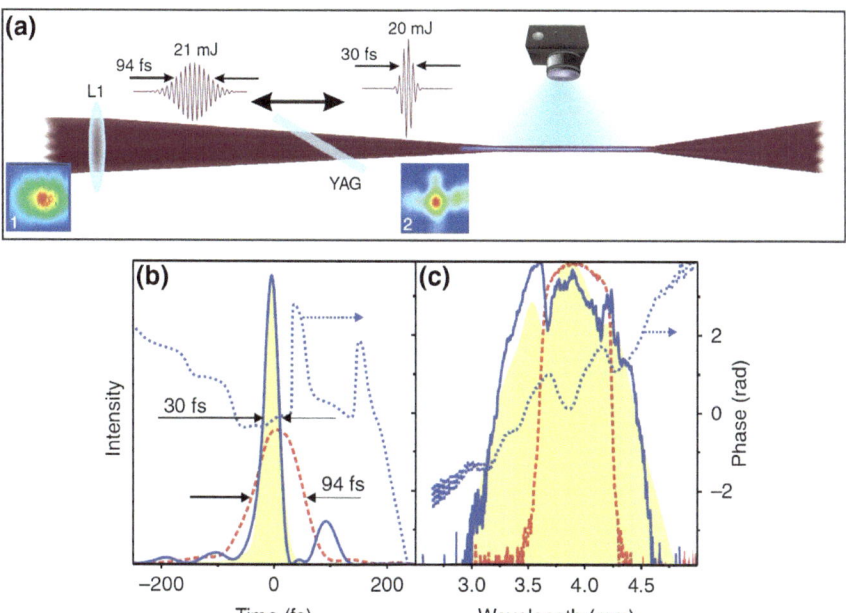

Fig. 6.5 a Sketch of the experimental setup illustrating a threefold self-compression of 94 fs, 21 mJ pulses at 3.9 mm, which is achieved in a 2-mm-thick YAG plate placed at a certain distance from the focusing lens L1. Insets 1 and 2 show the beam profiles on the focusing lens and on the output surface of the YAG plate, respectively. **b** temporal profiles of the input (dashed red curve) and self-compressed (blue solid curve) pulses retrieved from SHG-FROG measurements and **c** corresponding spectra. The dotted blue curves show retrieved temporal and spectral phases. Adapted from [52]

and carried a 0.44 TW peak power. The observed filamentation of the self-compressed pulses propagating in ambient air served as a convincing evidence for the increased peak power and high pulse and beam fidelity.

Using a similar focusing geometry, the soliton compression mechanism was demonstrated with systems of much simpler design based on optical parametric amplification. To this end, self-compression down to sub-three optical cycle widths of parametrically amplified difference frequency pulses in the $3-4$ µm wavelength range with several tens of µJ energy was demonstrated in YAG, CaF_2, and BaF_2 plates of millimetric thickness located before the geometric focus of a focusing lens [53]. Figure 6.6a shows the pulse width and energy transmission as functions of the position z of the CaF_2 plate with respect to the geometric focus ($z = 0$ mm), clearly distinguishing between self-compression regimes in the absence (lossless self-compression, far from the geometric focus) and in the presence (in the vicinity of the geometric focus) of filamentation. Figure 6.6b and c presents the spectrum, temporal profile, and phase of a 31 fs pulse that is self-compressed in a 4-mm-thick CaF_2 plate located at 6.5 mm before the geometric focus. A smooth spatial profile

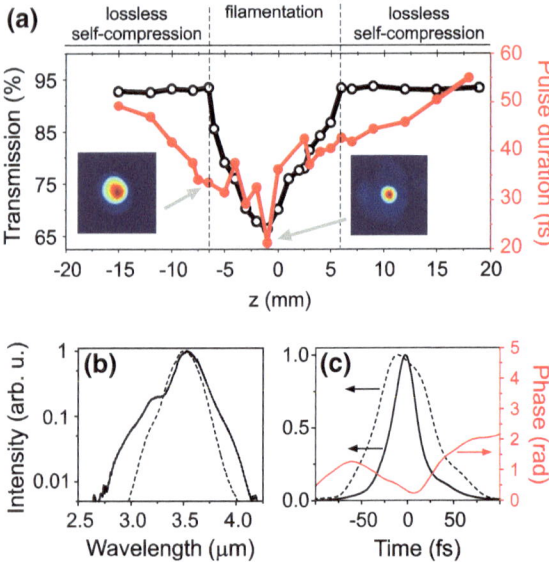

Fig. 6.6 Self-compression of 60 fs, 3.5 μm pulses in a 3-mm-thick CaF$_2$ plate. **a** Pulse duration (black curve) and energy transmission (red curve) as functions of the position z of the CaF$_2$ plate with respect to the geometric focus ($z = 0$ mm). The insets show the spatial intensity distributions of the self-compressed pulses in the absence ($z = -6.5$ mm) and the presence ($z = 0$ mm) of filamentation. **b** spectrum and **c** intensity profile and phase of 31 fs self-compressed pulse, retrieved from SFG-FROG measurements. The dashed curves show the spectrum and intensity profile of the input pulse. Adapted from [53]. Reprinted by permission from IOP Publishing

that is shown in the inset indicates the absence of filamentation. An energy through-put efficiency of above 90% was measured under these operational settings, yielding self-compressed pulses with sub-30 μJ energy, with the energy losses occurring just due to Fresnel reflections from the input and output faces of the nonlinear crystal. Fine management of the material nonlinearity and dispersion was demonstrated by compressing mid-infrared multi-cycle pulses (120 fs, 149 μJ, 3.5 μm) from a KTA-based optical parametric amplifier down to sub-two optical cycles (21 fs) using a combination of distributed YAG and Si plates and a pair of CaF$_2$ wedges with 83.5% energy throughput for the full beam [54]. A chirp-free pulse with a nearly Gaussian beam profile was obtained after additional spatial filtering, but at the expense of approximately twofold losses in energy transmission. The measured fluctuations of the carrier envelope phase stability of the self-compressed pulses were below 300 mrad (rms) attesting the robustness of that compression scheme.

The multi-plate setup that took advantage of soliton compression in the anoma-lous dispersion regime was used to self-compress 19 μJ, 63 fs, 1.55 μm pulses generated by an OPCPA pumped by a femtosecond ytterbium-doped fiber amplifier system, down to a 22 fs pulse duration with an energy transmission of 73% [55]. Here, the self-compression was performed in a 2-mm-thick anti-reflection coated fused silica plate placed asymmetrically between the curved mirrors, so as to retain temporal nonlinearities while homogenizing the spatial Kerr effect. The optical path was adjusted for the beam to make 10 round trips, resulting in an overall propagation distance in fused silica of 40 mm.

Due to technical simplicity and high-energy throughput, the soliton compression appears beneficial for the development of tabletop, high-power attosecond solid-state sources based on commercial laser technology. In that regard, self-compression of mid-infrared pulses at 3.8 μm from 90 fs to 35 fs in a 2-mm-thick YAG plate was demonstrated to increase the high-order harmonic yield by a factor of two and the cutoff energy from the 15th to beyond the 17th harmonic in a ZnO crystal [56].

6.2.5 Other Compression Mechanisms

Several other mechanisms for pulse compression were demonstrated so far, which either assist GVD-induced self-compression or rely on sole reshaping of intense laser pulses due to absorption and defocusing by the free electron plasma that is formed in the wake of the pulse. Pulse self-compression and spectral broadening was uncovered for the propagation of intense (with intensities of a few TW/cm^2) mid-infrared pulses in a thin piece of a nonlinear medium (a $\sim$2 mm CaF_2 plate) [57]. Here, self-compression of the driving pulse was shown to occur due to the generation of a free electron plasma, which absorbs and defocuses the trailing part of the pulse. The plasma-induced pulse compression resulted in spectral broadening around the carrier wavelength and facilitated large-scale spectral broadening of the third, fifth, and seventh harmonic spectra via cross-phase modulation, eventually yielding an SC spanning more than four octaves from the ultraviolet to the mid-infrared. Interestingly, the overlapping spectra of odd harmonics resulted in a remarkable spectral broadening into the deep ultraviolet, which extended well beyond the blue-shifted cutoff of a typical SC spectrum set by the material GVD. Plasma-induced self-compression along with broadband four-wave mixing was shown to dramatically broaden the pulse spectrum in the mid-infrared, as demonstrated in a GaAs crystal around its zero-GVD wavelength (6.8 μm) [58]. Supplementary post-compression of an octave-spanning spectrum in a 0.5-mm-thick BaF_2 plate yielded 20 fs pulse at the output that is less than 0.9 electric field cycles.

Theoretical investigation of filamentation with mid- and long-wavelength infrared laser pulses in a ZnSe crystal, which features anomalous GVD for the input wavelengths, predicted generation of light bullets with sub-cycle temporal width. Self-compression of the input pulses from 60 fs down to 15 fs at 8 μm and to 19 fs at 10 μm due to the collective action of extreme self-steepening, anomalous GVD, and plasma was demonstrated by numerical simulations [59].

6.3   Supercontinuum Generation with Picosecond Laser Pulses

SC generation with picosecond laser pulses in a solid-state medium is a challenging task, because of the competing effect of optical damage. Free electrons produced by multiphoton absorption at the leading part of the pulse are efficiently accelerated by the electric field of the trailing part of the pulse via the inverse Bremsstrahlung effect and acquire sufficient kinetic energy to initiate the avalanche, which eventually produces the optical breakdown of the medium. Therefore, a conventional and straightforward way for SC generation with picosecond laser pulses is to use liquids, such as water, where no permanent optical damage occurs. This approach was widely used in the past [60] and although seldom, it is still in use at present [61, 62]. SC generation in liquids was demonstrated to work equally well with few picosecond pulses from picosecond Ti:sapphire lasers [63], as well as with pulses of a few tens of picoseconds from Nd:YAG [64], mainly serving to produce the seed for picosecond (narrow band) optical parametric amplifiers.

The invention of the high-energy thin-disk oscillator and Innoslab laser amplifier technologies as well as the advent of large-mode fibers boosted the development of novel high repetition rate picosecond Yb-based solid-state laser sources emitting around 1 μm. The progress in picosecond laser technologies in turn prompted a renewed interest in SC generation with picosecond laser pulses in solid-state dielectric media. This interest is largely inspired by the demand of a robust broadband all-solid-state seeding source for compact, efficient, and inexpensive tabletop OPCPA systems that are solely based on this new generation of picosecond laser sources and provide few optical cycle pulses at the output.

The first studies aiming at stable SC generation with relatively long pulses in solids were performed with Ti:sapphire laser pulses of variable duration, produced via either spectral narrowing or pulse chirping [65, 66]. Owing to their high optical damage thresholds, sapphire and YAG were identified as the most suitable nonlinear materials for SC generation using sub-picosecond and picosecond laser pulses. The follow-up experiments using genuine sub-picosecond pulses from Yb:KYW [67, 68], Yb:YAG [69, 70], and 1-ps pulses from Nd:glass [71, 72] lasers demonstrated that under suitably chosen experimental conditions, a stable and reproducible SC could be generated without damaging the nonlinear material.

A detailed experimental and numerical investigation of SC generation with picosecond pulses revealed that the filamentation dynamics of a picosecond laser pulse in a normally dispersive medium shows a number of distinctive differences from filamentation dynamics of a femtosecond laser pulse [72]. These differences originate from a larger amount of free electron plasma, which is very efficiently produced by the long pulse around the nonlinear focus. As a result, a picosecond-pulsed laser beam undergoes a series of free electron plasma-driven transformations in space and time, which are distributed in the course of pulse propagation in the nonlinear material.

Fig. 6.7 Numerically simulated dynamics of **a** the beam radius and plasma density, and corresponding **b** temporal and **c** spectral evolutions of a 1.3-ps, 1055-nm, 5.1-μJ laser pulse versus propagation distance in a YAG crystal. The input pulse intensity is 300 GW/cm^2. Arrows indicate the nonlinear foci of the beam. The SC is produced after beam refocusing at the second nonlinear focus, where splitting of the replenished pulse occurs

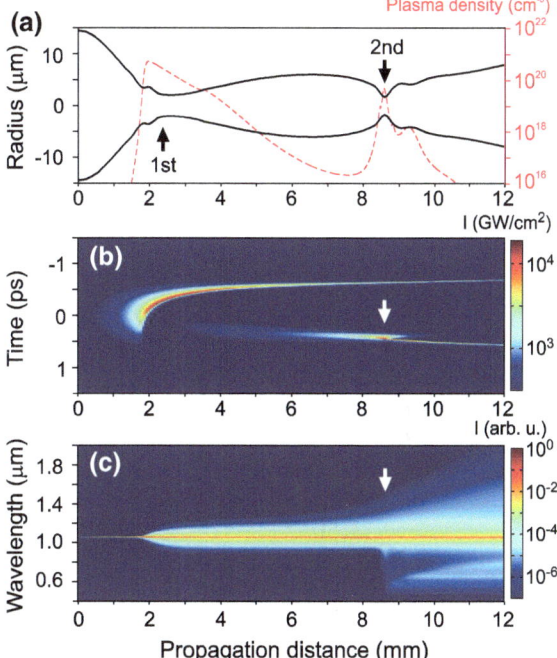

As an example, Fig. 6.7 illustrates the dynamics of the beam diameter and plasma density, and resulting evolutions of the temporal profile and spectrum of a 1.3 ps laser pulse with a carrier wavelength of 1055 nm versus propagation distance in a YAG crystal. In the vicinity of the (first) nonlinear focus, the free electron plasma, whose highest density peaks on the propagation axis, absorbs and defocuses the rear part of the pulse, inducing a considerable pulse shortening, and shifts the intensity peak to the front of the pulse. In the spectral domain, this induces only a rather moderate spectral broadening around the carrier frequency. As a result, no pulse splitting occurs at the first nonlinear focus, where a large fraction of the pulse energy is completely pushed out of the propagation axis, forming a ring-shaped intensity distribution at the tail. As soon as the plasma density ceases, this ring-shaped structure replenishes a short (femtosecond) pulse on the propagation axis. Thereafter, the replenished pulse self-focuses at the second nonlinear focus, where the pulse splitting occurs, which in turn produces the SC.

The uncovered picosecond pulse filamentation and SC generation scenario is therefore markedly different from femtosecond filamentation. The temporal transformations of the pulse at the nonlinear foci are in particular different since a femtosecond wave packet collapses as a whole, producing pulse splitting at the nonlinear focus and subsequent splittings with each refocusing cycle, as described in Sect. 4.4. Depending on the input pulse duration and energy, and on the external focusing geometry, plasma-driven transformations after the first nonlinear focus may result

Fig. 6.8 Supercontinuum
spectra in a 15-mm-long **a**
YAG and **b** sapphire
samples, as generated with
1055 nm, 1 ps pulses from a
Nd:glass laser, with energies
of 12 μJ and 21 μJ,
respectively. Insets show the
screenshots of the SC
emission patterns in the far
field. Reprinted from [71]
with permission

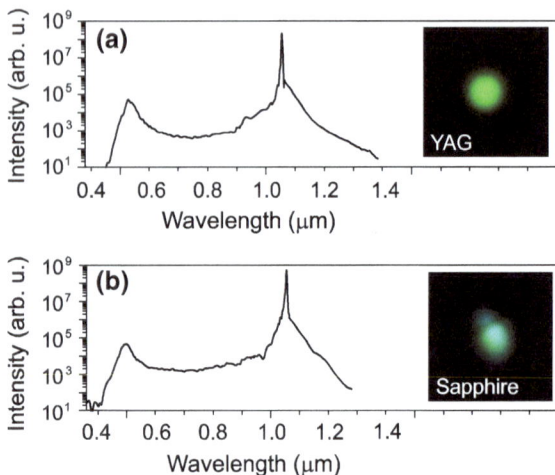

in a more complex evolution of the pulse versus propagation distance, and hence
in a multi-peaked temporal structure at the output of the nonlinear medium, see,
e.g., [72].

A more recent study uncovered that filamentation of relatively long femtosecond
pulses, having pulse widths of a few hundreds of fs shares many common features with the picosecond pulse filamentation dynamics and results in an interesting,
if not surprising dynamics of spectral broadening, especially when focused inside
the nonlinear material [73]. These findings should be considered for building and
optimization of SC generation schemes, where SC is produced by relatively long
femtosecond and sub-picosecond pulses delivered by Yb:KGW, Yb:KYW, Yb:fiber,
and similar laser systems.

As a result of complex temporal dynamics, SC generation with picosecond laser
pulses requires longer samples of the nonlinear media, which may vary from ∼10
mm to as much as 15 cm, depending on the focusing condition of the input beam and
pulse width. Despite the aforementioned differences, picosecond laser pulses produce
the SC spectra, whose extent is nearly identical to that produced with femtosecond
pulses. However, picosecond SC exhibits a remarkably higher spectral intensity at
the vicinity of the carrier wavelength, see Fig. 6.8, which in many measurements is
usually filtered out by using highly reflective mirrors for the pump wavelength.

The problem of optical damage becomes more severe using pump pulses with
shorter carrier wavelengths. To this end, SC generation was reported with sub-
picosecond pulses at 515 nm (second harmonic of an Yb:KYW laser) in sapphire
and YAG, which produced SC spectra covering the 400–800 nm wavelength range,
but the optical damage-free operation was achieved only for a very narrow range of
focusing conditions [67]. Quite similar results were reported in a YAG crystal using
515 nm, 466 fs pulses, generated by frequency doubling and cross-polarized wave
generation at the output of a 1-ps Yb:YAG regenerative amplifier [74].

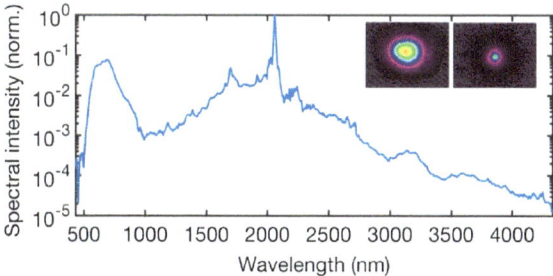

Fig. 6.9 SC spectrum generated in a 15 cm YAG crystal with 3 ps, 150 µJ pulses at 2.05 µm with low numerical aperture focusing geometry. The inset shows the spatial profiles of the pump (left) and the SC (right) beams in the far field. Adapted from [76]. Reprinted by permission from the Optical Society of America

A couple of SC generation experiments were performed with pump pulses of a few picoseconds (3 ps) and very long (up to 15 cm) YAG slabs in loose focusing geometry. In that regard, stable SC generation was reported with slightly chirped 3 ps pulses at 1030 nm from an Yb:YAG thin-disk regenerative amplifier [75]. The blue-shifted portion of SC was fully characterized and demonstrated to allow compression of the broadband pulse down to 15 fs.

An ultrabroadband SC (in the 0.5−4.5 µm range) was generated in a 15-cm-long YAG crystal using 3 ps pulses with a carrier wavelength of 2.05 µm from a Ho:YLF regenerative amplifier, [76], see Fig. 6.9. Loose focusing of the input beam (NA = 0.005) coupled with a crystal length so as to nearly match the Rayleigh range yielded stable SC generation with the long-term stability of the visible-near-infrared and mid-infrared parts of the spectrum being almost the same as the pump laser, while SHG-FROG measurements revealed a complex, multi-peaked temporal profile of the SC pulse.

High temporal coherence of the SC generated by sub-picosecond pulses was directly verified by the spectral phase measurements of the parametrically amplified portion of the SC spectrum [67]. More recently, the generation of a very stable and compressible SC was demonstrated with a Yb:YAG laser system based on the InnoSlab concept, which delivers multimilijoule picosecond pulses with high average power exceeding half of a kW [77]. Figure 6.10a shows the blue-shifted portion of the SC spectrum generated with 1.03 ps, 1030 nm, 2.5 µJ pulses in a YAG crystal at a repetition rate of 20 kHz. The generated SC shows excellent long-term reproducibility of the spectral shape and stability of the energy. The measured variations of the RMS noise of the integrated SC power in the 500 − 1000 nm range fall between 0.1 and 0.3% over 3 h of operation, as illustrated in Fig. 6.10b, with the RMS noise of the pump laser of 0.25%. Post-compression of the SC spectral parts around 700 nm and 600 nm enables a 100-fold compression down to 10 fs and 14 fs, respectively, using a conventional reflection grating setup.

The availability of picosecond pulse-generated SC with well-behaved spectral phase that is compressible down to the transform limit paved new avenues for

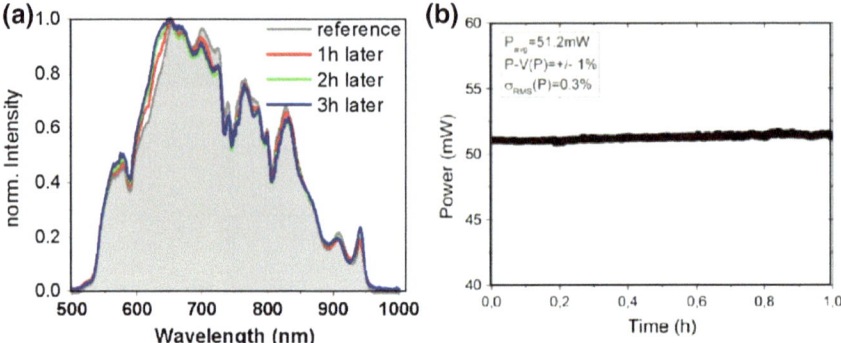

Fig. 6.10 **a** Stability of the blue-shifted portion of the SC spectrum generated with 1.03 ps, 1030 nm, 2.5 µJ pulses in a YAG crystal, characterized over 3 h of operation at a repetition rate of 20 kHz. **b** The RMS noise of the integrated supercontinuum power in the 500–1000 nm range. Adapted from [77]. Reprinted by permission from the Optical Society of America

the development of picosecond laser pumped and SC-seeded noncollinear optical parametric amplifiers [74, 78] and of a whole new generation of compact OPCPA systems. These OPCPA systems are built around the recently developed amplified sub-picosecond and picosecond lasers, such as Yb:KYW [68, 79], Yb:YAG [69, 70, 75, 80, 81], Yb-fiber [82, 83], and mixed laser systems, which combine Yb:fiber oscillators with Yb:YAG regenerative amplifiers [84] and are designed to provide few optical cycle pulses in the near- and mid-infrared at very high repetition rates. Picosecond SC is also employed as a seed signal for the mid-infrared optical parametric amplifiers that are driven by picosecond Ho:YAG amplifiers and provide few optical cycle mid-infrared pulses with carrier wavelengths around 5 µm [85, 86].

6.4   Control of Supercontinuum Generation

The rapidly expanding field of applications calls for achieving broadband radiation with desired temporal and spectral properties, which in turn require setting an efficient control on the femtosecond filamentation process. The filamentation dynamics and hence the spectral content of the SC are defined essentially by the laser wavelength and by linear and nonlinear parameters of the medium, such as bandgap, nonlinear index of refraction, and chromatic dispersion. These material parameters possess fundamental mutual relationships and therefore are generally fixed for a given nonlinear medium at a given incident wavelength, greatly restraining the possibilities to influence the filamentation dynamics and the process of spectral broadening in real experimental settings.

   Several conceptually different approaches have been proposed to overcome these limitations, allowing for changes or modifications in the spectral content of the SC radiation in a controlled fashion. The first approach relies on the modification and tailoring of the incident pulse and beam parameters. A certain degree of control of spectral broadening was achieved by relatively simple means. To this end, control of filamentation and SC generation processes was demonstrated by varying the polarization state (from linear to circular) of the incident pulse in isotropic media [87, 88]. In a birefringent medium with vanishing second-order susceptibility, e.g., an α-BBO crystal, the input pulses of either ordinary or extraordinary polarizations were shown to produce different SC spectra owing to different nonlinear indexes of refraction for the $o-$ and $e-$polarized waves [89]. Setting of either positive or negative chirp of the input pulse allowed controllable tuning of the wavelength of the blue-shifted cutoff of the SC spectrum [90]. Similar results were achieved by the adjustment of the focal plane of the incident beam [91, 92]. An enhancement of the SC generation induced by Fresnel diffraction of the input beam from a circular aperture was reported as well [93]. A considerable enhancement of the red-shifted portion of the SC spectrum was achieved by setting low numerical aperture of the input beam and using a longer nonlinear medium [94, 95], see Sect. 4.2 for more details.

More sophisticated methods are based on proper phase and amplitude shaping of the input pulse and beam. Considerable (one order of magnitude) enhancement of the SC spectral intensity within specified bandwidths was achieved by shaping the amplitude and phase of the input pulse [96]. By introducing second- and third-order phase distortions controlled by an acousto-optic programmable dispersive filter, precise tailoring of the input pulse was demonstrated to affect the pulse splitting dynamics, which in turn resulted in the generation of an SC with controllable spectral bandwidth and shape [97]. More recently, a similar approach based on employing a 4-f pulse shaper with a phase mask consisting of a liquid crystal spatial light modulator was demonstrated for achieving accumulation of the SC spectral intensity at selected frequencies by a controllable amount [98]. In particular, it was demonstrated how the wavelength of the spectral peak generated by the amplitude concentration moves according to the spatial frequency of light being modulated, i.e., by simply scanning the modulating pixels across the spatial light modulator. The experimental results presented in Fig. 6.11 also demonstrate a possibility to produce multiple peaks in the SC spectrum, which separate according to the modulating pixel separation. More generally, these results suggest that a narrow-bandwidth laser pulse can be made tunable to any wavelength within the limits of the original bandwidth of the incident femtosecond laser pulse.

Control of the SC spectrum to a certain degree was also performed by changing the relative position of the focus in the nonlinear medium by means of diffractive optics [99, 100] or by varying the spatial phase of the input beam by means of programmable spatial light modulators [101–103]. Fine adjustment of the position of the nonlinear focus was demonstrated by varying the carrier-envelope phase of a few-cycle input pulse [104].

An efficient control of the SC spectral width and notable enhancement of the SC components within certain spectral ranges in particular was demonstrated by means of the so-called two-color filamentation, i.e., by launching two filament-forming beams of different wavelengths [105]. This configuration provides an additional

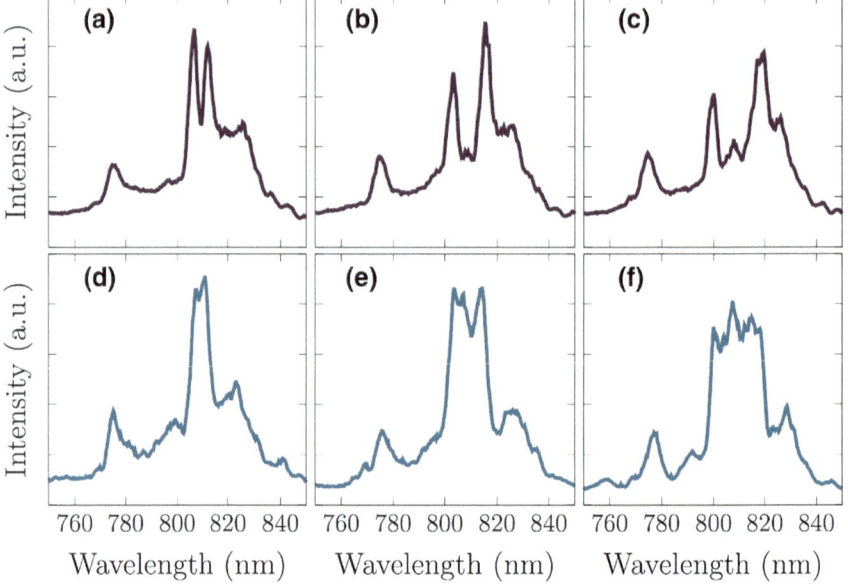

Fig. 6.11 Spectra of shaped filament in water for different phase mask configurations. Here, two narrow intensity spikes (top row) are formed or one wide intensity spike (bottom row) is formed by correctly choosing which pixels of the spatial light modulator to modulate. Adapted from [98]

degree of freedom in controlling the SC generation process via the time delay between the pump pulses, as demonstrated in the geometry of crossing beams, in the single [106] and to some extent, in the multiple [107] filamentation regimes. More recently, control of the polarization of the SC spectrum was demonstrated by changing the angle between the polarization directions of two femtosecond pump pulses in a piece of K9 glass, which is optically isotropic [108].

A series of interesting results regarding modifications of the SC spectrum were obtained by collinear two-color filamentation in a sapphire crystal, when launching 300 fs pulses at the fundamental (1030 nm) and second harmonic (515 nm) wavelengths from an amplified Yb:KGW laser. A complex spectral behavior was captured by varying the time delay between the incident pulses, in particular demonstrating how certain bands in the SC spectrum could be either enhanced or completely suppressed. Figure 6.12a shows examples of the resulting SC spectra recorded by varying the time delay between the co-filamenting fundamental and second harmonic pulses. The individual SC spectra generated by filamentation of separately launched fundamental and second harmonic pulses serve as a reference and are presented in Fig. 6.12b.

The second approach is based on tailoring the linear (e.g., absorption) and nonlinear (nonlinear index of refraction) parameters of the medium itself. As suggested by the pioneering work [110], addition of various dopants may lead to an enhanced nonlinear optical response of the medium and thus may lower the threshold for

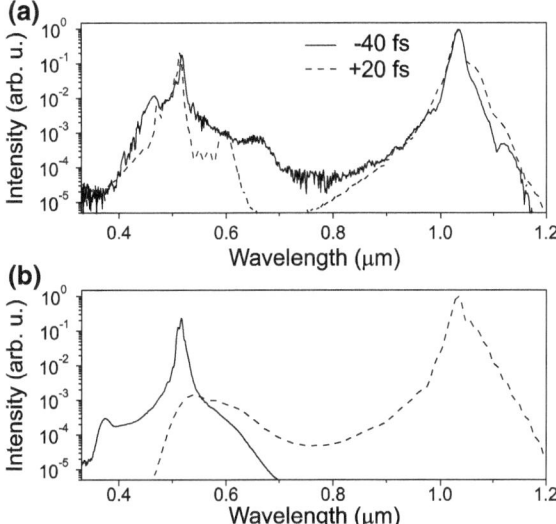

Fig. 6.12 **a** Resulting SC spectra generated in a 3-mm-long sapphire plate by co-filamentation of 300 fs pulses at 1030 nm and 515 nm, each containing a power equal to $2.6\ P_{cr}$. The negative delay time indicates that the second harmonic pulse comes first. **b** Individual SC spectra generated by launching the two input pulses separately. Adapted from [109] with permission

nonlinear processes. These early ideas received further attention concerning femtosecond filamentation and SC generation in water doped by silver [111] and gold [112] nanoparticles and gold nanorods [113], and in LN crystal doped with Ni ions [114]. These experiments demonstrated that an easy control of the threshold energy and spectral extent and shape of femtosecond supercontinuum may be achieved by varying the size and the concentration of dopants. Filamentation and SC generation with femtosecond laser pulses in aqueous salt solutions was studied as well [115], in particular, demonstrating that addition of absorbing inorganic dopants yields more flat SC spectra [116]. Significant modifications of SC spectra and even suppression of SC generation were demonstrated in water by adding protein [117, 118], lactose and nitric acid solutions [119], hence allowing the development of new diagnostic tools for chemical and biological systems. Femtosecond SC generation was also reported in more complex nonlinear media such as aqueous colloids containing silver nanoparticles [120] and nanocomposite materials [121]. Finally, it was demonstrated that the presence of neutral scatterers in a liquid medium significantly affects SC generation and alters its spectral shape. Significant suppression of the short-wavelength side of the SC spectrum was reported in water containing suspended polystyrene microspheres with a diameter ranging from 0.3 to 3 μm [122]. The SC suppression was attributed to the dissipation of the incident energy, strong scattering of the shorter wavelengths, and stretching of the pulse temporally.

Experimental studies of SC generation in laser-modified media revealed dramatic changes in SC spectra compared to those generated in unmodified materials. A significant narrowing of SC spectra and even a complete suppression of SC generation was observed in laser-modified fused silica, due to the rapidly evolving permanent change of refractive index induced by irradiation of the sample at very high repetition rates (up to several hundreds of kHz) [123]. Laser-induced structural modifications of YAG produce peculiar SC spectra with strongly suppressed emission in the infrared [71].

Fig. 6.13 Supercontinuum spectra in a 5-mm-thick β-BBO crystal generated with 100 fs, 290 nJ input pulses at 800 nm and measured at various crystal detunings from the perfect phase-matching angle (29.2°): **a** 32.4°, **b** 37.5°, **c** 42.6°. The narrow peak at 400 nm and the broad peak with tunable wavelength are the phase-mismatched and self-phase-matched second harmonics, respectively. The insets show the visual appearances of the SC beam in the far field

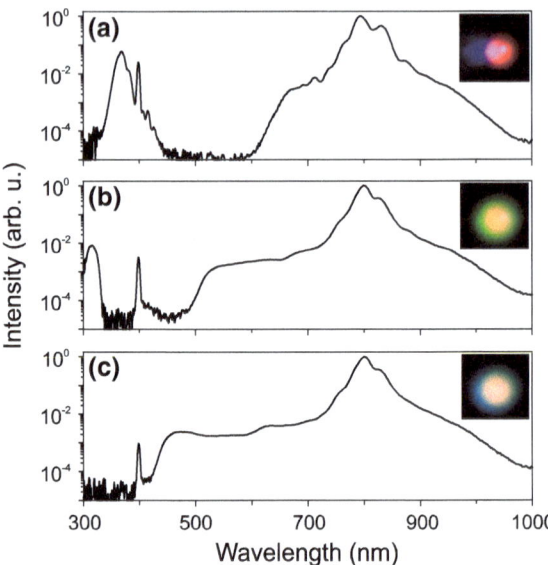

Finally, very simple and efficient, but still poorly explored possibilities regarding a full control of the SC spectral width were demonstrated in birefringent crystals by making use of second-order cascading effects, which were discussed in more detail in Sect. 5.5. Figure 6.13 shows the control of the blue-shifted portion of SC spectrum produced by filamentation of 90 fs, 800 nm laser pulses in a β-BBO crystal, as the crystal is rotated out of perfect phase matching for second harmonic generation ($\theta = 29.2°$), in the angle range that provides enhanced self-focusing [124], see Fig. 5.10 of Sect. 5.5. The achieved control of the SC spectrum was demonstrated to be very robust: the blue-shifted cutoff wavelength was remarkably stable for a given crystal orientation and did not change by varying the input pulse energy in rather broad confines, as long as the single filamentation regime is preserved. Such a robust spectral control was attributed to the generation of the self-phase matched second harmonic, which emerges as a broad blue-shifted and cross-polarized spectral peak with a tunable central wavelength, as shown in Fig. 6.13a and b. The generation of the self-phase matched second harmonic introduces a considerable energy loss only to the trailing pulse after pulse splitting, thus counteracting the effect of self-steepening, which gives rise to the blue-shifted spectral broadening.

6.5 Supercontinuum Generation with Non-Gaussian Beams

The vast majority of the experiments on SC generation were performed using conventional, Gaussian-shaped input beams. However, several studies were devoted to investigate the SC generation with more complex input waveforms. In that regard, femtosecond Bessel beams are of particular interest owing to their non-diffractive property, which ensures high-energy localization within the intense central spot over long distances, well exceeding the Rayleigh range of conventional Gaussian beams with the same spot dimensions. In addition, there is a great flexibility of producing a Bessel beam with desired properties: cone angle, central spot size, energy content, particular intensity variation along the propagation axis, and length of the Bessel zone, which can be adjusted by relatively simple means, e.g., by Bessel beam formation geometry. SC generation using femtosecond Bessel beams with various parameters was studied in liquids: water [125] and methanol [126], and in solids: sapphire [127] and BaF$_2$ [128, 129]. These studies revealed that filamentation of a femtosecond Bessel beam shows a number of differences compared with the case of Gaussian beam filamentation, and these differences result in a different scenario for spectral broadening and SC generation, in particular, demonstrating that the SC radiation emerges on the propagation axis, as illustrated in Fig.6.14. On the other hand, due to the conical energy flow and energy distribution between the central spot and surrounding rings, Bessel beams are very resistant to self-focusing, and typically, a 10−100 times higher incident energy is required to induce filamentation and achieve appreciable spectral broadening.

A couple of studies were devoted to investigate spectral broadening and SC generation with more sophisticated waveforms. Filamentation and SC generation experiments with self-bending Airy beams in water [130, 131] and in fused silica [132] had shown that the interpretation of SC emission patterns may provide quantitative clues on the complex evolution of these sophisticated waveforms in the highly nonlinear propagation regime. Finally, SC generation was investigated in CaF$_2$ with vortex [133] and singular [134] beams, which are characterized by the presence of spiral and step-wise phase dislocations in the beam wavefront, respectively, and which determine its phase and intensity structure. In both cases, it was shown that the generated SC experiences large divergence and appears as a wide white-light back-

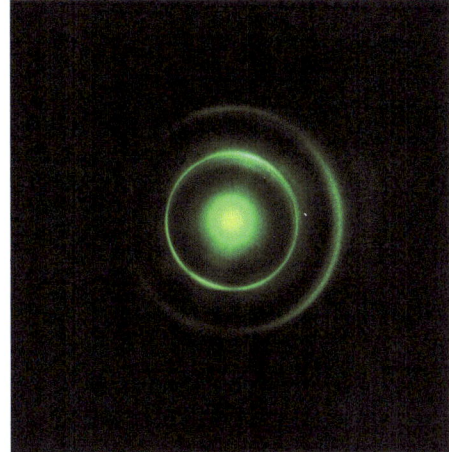

Fig. 6.14 Far-field image of the SC in water generated by filamentation of an axicon-focused Bessel beam carrying 200 fs pulses with a central wavelength of 527 nm. The SC beam emerges as a bright spot in the center. The inner ring corresponds to the input Bessel beam, while the outer ring emerges due to four-wave mixing between the axial and conical components

ground surrounding the original incident beam, whose spatial profile remains well preserved in the process of SC generation.

6.6 Other Developments

Several methods based on broadband frequency conversion in crystals with second-order nonlinearity were demonstrated aiming at broadband frequency conversion of the SC radiation. These include frequency up-conversion of the SC radiation into spectral regions that are difficult to access via direct SC generation, e.g., ultraviolet, and frequency down-conversion of the broadband radiation with desired shaping of its spectral and temporal characteristics, in the pursuit of ultimate control of single-cycle pulses.

Ultraviolet is a highly desired, but still experimentally very hardly accessible spectral region for SC generation. Generally, filamentation of ultraviolet laser pulses in wide bandgap solid-state dielectric materials produces only rather moderate spectral broadening at best, see Chap. 4. Spectral broadening is constrained by material dispersion, which rapidly increases toward higher frequencies, while due to large incident photon energy and low order of the nonlinear absorption, all materials are prone to rapid degradation and optical damage. To overcome these fundamental limitations, an interesting approach was proposed based on frequency up-conversion exploiting greatly relaxed phase-matching conditions supported by random quasi-phase matching in polycrystalline media with second-order nonlinearity. The visible portion of femtosecond supercontinuum generated in a photonic crystal fiber was converted into the ultraviolet (from 260 to 305 nm) via randomly quasi-phase-matched second harmonic generation in a polycrystalline strontium tetraborate (SBO) crystal, which exhibits high transparency in the vacuum ultraviolet down to 121 nm [135]. The demonstrated methodology may appear particularly beneficial for frequency up-conversion of the SC generated in bulk nonlinear media, since it has a considerably higher spectral energy density compared to that extracted from optical fibers, especially bearing in mind energy scaling capabilities of the SC radiation making use of the multi-plate technique.

Adiabatic frequency conversion is a powerful method that enables to overcome bandwidth limitations of three wave-mixing schemes [136]. Recently, adiabatic difference frequency generation using a broadband chirped near-infrared pulse in an aperiodically poled magnesium-oxide-doped congruent lithium niobate (MgO:CLN) was demonstrated to produce a single-cycle pulse with a duration of 11 fs and a multi-octave-spanning spectrum in the mid-infrared, in the $1.8-4.4$ μm range [137]. The nature of the adiabatic frequency conversion process allows for a one-to-one transfer of the spectral phase through nonlinear frequency conversion over an extremely large spectral range. This unique feature, combined with the possibility to adjust the amplitude and phase of the near-infrared pulse prior to conversion in a desired way, offers an unprecedented capability to shape the spectral and temporal characteristics of down-converted pulses over their entire bandwidth. It is foreseen that nonlinear materials such as orientation-patterned semiconductors may offer a route to extending the conversion bandwidth by multiple octaves in the mid-wave and long-wave infrared spectral ranges.

References

1. Lu, C.-H., Tsou, Y.-J., Chen, H.-Y., Chen, B.-H., Cheng, Y.-C., Yang, S.-D., Chen, M.-C., Hsu, C.-C., Kung, A.H.: Generation of intense supercontinuum in condensed media. Optica **1**, 400–406 (2014)
2. Cheng, Y.-C., Lu, C.-H., Lin, Y.-Y., Kung, A.H.: Supercontinuum generation in a multi-plate medium. Opt. Express **24**, 7224–7231 (2016)
3. He, P., Liu, Y., Zhao, K., Teng, H., He, X., Huang, P., Huang, H., Zhong, S., Jiang, Y., Fang, S., Hou, X., Wei, Z.: High-efficiency supercontinuum generation in solid thin plates at 0.1 TW level. Opt. Lett. **42**, 474–477 (2017)
4. Budriūnas, R., Kučinskas, D., Varanavičius, A.: High-energy continuum generation in an array of thin plates pumped by tunable femtosecond IR pulses. Appl. Phys. B **123**, 212 (2017)
5. Shank, C.V., Fork, R.L., Yen, R., Stolen, R.H., Tomlinson, W.J.: Compression of femtosecond optical pulses. Appl. Phys. Lett. **40**, 761–763 (1982)
6. Tomlinson, W.J., Stolen, R.H., Shank, C.V.: Compression of optical pulses chirped by self-phase modulation in fibers. J. Opt. Soc. Am. B **1**, 139–149 (1984)
7. Fork, R.L., Brito Cruz, C.H., Becker, P.C., Shank, C.V.: Compression of optical pulses to six femtoseconds by using cubic phase compensation. Opt. Lett. **12**, 483–485 (1987)
8. Nisoli, M., De Silvestri, S., Svelto, O.: Generation of high energy 10 fs pulses by a new pulse compression technique. Appl. Phys. Lett. **68**, 2793–2795 (1996)
9. Cardin, V., Thiré, N., Beaulieu, S., Wanie, V., Légaré, F., Schmidt, B.E.: 0.42 TW 2-cycle pulses at 1.8 μm via hollow-core fiber compression. Appl. Phys. Lett. **107**, 181101 (2015)
10. Balciunas, T., Fourcade-Dutin, C., Fan, G., Witting, T., Voronin, A.A., Zheltikov, A.M., Gerome, F., Paulus, G.G., Baltuska, A., Benabid, F.: A strong-field driver in the single-cycle regime based on self-compression in a kagome fibre. Nat. Commun. **6**, 6117 (2015)
11. Couairon, A., Biegert, J., Hauri, C.P., Kornelis, W., Helbing, F.W., Keller, U., Mysyrowicz, A.: Self-compression of ultra-short laser pulses down to one optical cycle by filamentation. J. Mod. Opt. **53**, 75–85 (2006)
12. Mitrofanov, A.V., Voronin, A.A., Rozhko, M.V., Sidorov-Biryukov, D.A., Fedotov, A.B., Pugžlys, A., Shumakova, V., Ališauskas, S., Baltuška, A., Zheltikov, A.M.: Self-compression of high-peak-power mid-infrared pulses in anomalously dispersive air. Optica **4**, 1405–1408 (2017)
13. Rolland, C., Corkum, P.B.: Compression of high-power optical pulses. J. Opt. Soc. Am. B **5**, 641–647 (1988)
14. Mével, E., Tcherbakoff, O., Salin, F., Constant, E.: Extracavity compression technique for high-energy femtosecond pulses. J. Opt. Soc. Am. B **20**, 105–108 (2003)
15. Seidel, M., Arisholm, G., Brons, J., Pervak, V., Pronin, O.: All solid-state spectral broadening: an average and peak power scalable method for compression of ultrashort pulses. Opt. Express **24**, 9412–9428 (2016)
16. Pronin, O., Seidel, M., Lücking, F., Brons, J., Fedulova, E., Trubetskov, M., Pervak, V., Apolonski, A., Udem, T., Krausz, F.: High-power multi-megahertz source of waveform-stabilized few-cycle light. Nat. Commun. **6**, 6998 (2015)
17. Krebs, N., Pugliesi, I., Riedle, E.: Pulse compression of ultrashort UV pulses by self-phase modulation in bulk material. Appl. Sci. **3**, 153–167 (2013)
18. Lanin, A.A., Voronin, A.A., Stepanov, E.A., Fedotov, A.B., Zheltikov, A.M.: Frequency-tunable sub-two-cycle 60-MW-peak-power free-space waveforms in the mid-infrared. Opt. Lett. **39**, 6430–6433 (2014)

19. Schulte, J., Sartorius, T., Weitenberg, J., Vernaleken, A., Russbueldt, P.: Nonlinear pulse compression in a multi-pass cell. Opt. Lett. **41**, 4511–4514 (2016)
20. Beetar, J.E., Gholam-Mirzaei, S., Chini, M.: Spectral broadening and pulse compression of a 400 μJ, 20 W Yb:KGW laser using a multi-plate medium. Appl. Phys. Lett. **112**, 051102 (2018)
21. Fritsch, K., Poetzlberger, M., Pervak, V., Brons, J., Pronin, O.: All-solid-state multipass spectral broadening to sub-20 fs. Opt. Lett. **43**, 4643–4646 (2018)
22. Lu, C.-H., Witting, T., Husakou, A., Vrakking, M.J.J., Kung, A.H., Furch, F.J.: Sub-4 fs laser pulses at high average power and high repetition rate from an all-solid-state setup. Opt. Express **26**, 8941–8956 (2018)
23. Voronin, A.A., Zheltikov, A.M., Ditmire, T., Rus, B., Korn, G.: Subexawatt few-cycle light wave generation via multipetawatt pulse compression. Opt. Commun. **291**, 299–303 (2013)
24. Farinella, D.M., Wheeler, J., Hussein, A.E., Nees, J., Stanfield, M., Beier, N., Ma, Y., Cojocaru, G., Ungureanu, R., Pittman, M., Demailly, J., Baynard, E., Fabbri, R., Masruri, M., Secareanu, R., Naziru, A., Dabu, R., Maksimchuk, A., Krushelnick, K., Ros, D., Mourou, G., Tajima, T., Dollar, F.: Focusability of laser pulses at petawatt transport intensities in thin film compression. J. Opt. Soc. Am. B **36**, A28–A32 (2019)
25. Hache, F., Zéboulon, A., Gallot, G., Gale, G.M.: Cascaded second-order effects in the femtosecond regime in β-barium borate: self-compression in a visible femtosecond optical parametric oscillator. Opt. Lett. **20**, 1556–1558 (1995)
26. Liu, X., Qian, L., Wise, F.: High-energy pulse compression by use of negative phase shifts produced by the cascade $\chi^{(2)} : \chi^{(2)}$ nonlinearity. Opt. Lett. **23**, 1777–1779 (1999)
27. Ashihara, S., Nishina, J., Shimura, T., Kuroda, K.: Soliton compression of femtosecond pulses in quadratic media. J. Opt. Soc. Am. B **19**, 2505–2510 (2002)
28. Moses, J., Wise, F.W.: Soliton compression in quadratic media: high-energy few-cycle pulses with a frequency-doubling crystal. Opt. Lett. **31**, 1881–1883 (2006)
29. Zhou, B.B., Chong, A., Wise, F.W., Bache, M.: Ultrafast and octave-spanning optical nonlinearities from strongly phase-mismatched quadratic interactions. Phys. Rev. Lett. **109**, 043902 (2012)
30. Viotti, A.-L., Lindberg, R., Zukauskas, A., Budriunas, R., Kucinskas, D., Stanislauskas, T., Laurell, F., Pasiskevicius, V.: Supercontinuum generation and soliton self-compression in $\chi^{(2)}$-structured KTiOPO$_4$. Optica **5**, 711–717 (2018)
31. Seidel, M., Brons, J., Arisholm, G., Fritsch, K., Pervak, V., Pronin, O.: Efficient high-power ultrashort pulse compression in self-defocusing bulk media. Sci. Rep. **7**, 1410 (2017)
32. Zhou, B., Guo, H., Bache, M.: Energetic mid-IR femtosecond pulse generation by self-defocusing soliton-induced dispersive waves in a bulk quadratic nonlinear crystal. Opt. Express **23**, 6924–6936 (2015)
33. Zhou, B., Bache, M.: Dispersive waves induced by self-defocusing temporal solitons in a beta-barium-borate crystal. Opt. Lett. **40**, 4257–4260 (2015)
34. Bache, M., Guo, H., Zhou, B.: Generating mid-IR octave-spanning supercontinua and few-cycle pulses with solitons in phase-mismatched quadratic nonlinear crystals. Opt. Mater. Express **3**, 1647–1657 (2013)
35. Šuminas, R., Tamošauskas, G., Dubietis, A.: Filamentation-free self-compression of mid-infrared pulses in birefringent crystals with second-order cascading-enhanced self-focusing nonlinearity. Opt. Lett. **43**, 235–238 (2018)
36. Chekalin, S.V., Kompanets, V.O., Smetanina, E.O., Kandidov, V.P.: Light bullets and supercontinuum spectrum during femtosecond pulse filamentation under conditions of anomalous group-velocity dispersion in fused silica. Quantum Electron. **43**, 326–331 (2013)
37. Durand, M., Jarnac, A., Houard, A., Liu, Y., Grabielle, S., Forget, N., Durécu, A., Couairon, A., Mysyrowicz, A.: Self-guided propagation of ultrashort laser pulses in the anomalous dispersion region of transparent solids: a new regime of filamentation. Phys. Rev. Lett. **110**, 115003 (2013)
38. Gražulevičiūtė, I., Šuminas, R., Tamošauskas, G., Couairon, A., Dubietis, A.: Carrier-envelope phase-stable spatiotemporal light bullets. Opt. Lett. **40**, 3719–3722 (2015)

39. Chekalin, S.V., Dokukina, A.E., Dormidonov, A.E., Kompanets, V.O., Smetanina, E.O., Kandidov, V.P.: Light bullets from a femtosecond filament. J. Phys. B **48**, 094008 (2015)
40. Majus, D., Tamošauskas, G., Gražulevičiūtė, I., Garejev, N., Lotti, A., Couairon, A., Faccio, D., Dubietis, A.: Nature of spatiotemporal light bullets in bulk Kerr media. Phys. Rev. Lett. **112**, 193901 (2014)
41. Hemmer, M., Baudisch, M., Thai, A., Couairon, A., Biegert, J.: Self-compression to sub-3-cycle duration of mid-infrared optical pulses in dielectrics. Opt. Express **21**, 28095–28102 (2013)
42. Baudisch, M., Pires, H., Ishizuki, H., Taira, T., Hemmer, M., Biegert, J.: Sub-4-optical-cycle, 340 MW peak power, high stability mid-IR source at 160 kHz. J. Opt. **17**, 094002 (2015)
43. Liang, H., Krogen, P., Grynko, R., Novak, O., Chang, C.-L., Stein, G.J., Weerawarne, D., Shim, B., Kärtner, F.X., Hong, K.-H.: Three-octave-spanning supercontinuum generation and sub-two-cycle self-compression of mid-infrared filaments in dielectrics. Opt. Lett. **40**, 1069–1072 (2015)
44. Chekalin, S.V., Kompanets, V.O., Dormidonov, A.E., Kandidov, V.P.: Path length and spectrum of single-cycle mid-IR light bullets in transparent dielectrics. Quantum Electron. **48**, 372–377 (2018)
45. Šuminas, R., Tamošauskas, G., Valiulis, G., Dubietis, A.: Spatiotemporal light bullets and supercontinuum generation in β-BBO crystal with competing quadratic and cubic nonlinearities. Opt. Lett. **41**, 2097–2100 (2016)
46. Voronin, A.A., Zheltikov, A.M.: Asymptotically one-dimensional dynamics of high-peak-power ultrashort laser pulses. J. Opt. **18**, 115501 (2016)
47. Voronin, A.A., Zheltikov, A.M.: Pulse self-compression to single-cycle pulse widths a few decades above the self-focusing threshold. Phys. Rev. A **94**, 023824 (2016)
48. Balakin, A.A., Kim, A.V., Litvak, A.G., Mironov, V.A., Skobelev, S.A.: Extreme self-compression of laser pulses in the self-focusing mode resistant to transverse instability. Phys. Rev. A **94**, 043812 (2016)
49. Bravy, B.G., Gordienko, V.M., Platonenko, V.T.: Kerr effect-assisted self-compression in dielectric to single-cycle pulse width and to terawatt power level in mid-IR. Opt. Commun. **344**, 7–11 (2015)
50. Bravy, B.G., Gordienko, V.M., Platonenko, V.T.: Self-compression of terawatt level picosecond 10 μm laser pulses in NaCl. Laser Phys. Lett. **11**, 065401 (2014)
51. Li, W., Li, Y., Xu, Y., Guo, X., Lu, J., Wang, P., Leng, Y.: Design and simulation of a single-cycle source tunable from 2 to 10 micrometers. Opt. Express **25**, 7101–7111 (2017)
52. Shumakova, V., Malevich, P., Ališauskas, S., Voronin, A., Zheltikov, A.M., Faccio, D., Kartashov, D., Baltuška, A., Pugžlys, A.: Multi-millijoule few-cycle mid-infrared pulses through nonlinear self-compression in bulk. Nat. Commun. **7**, 12877 (2016)
53. Marcinkevičiūtė, A., Garejev, N., Šuminas, R., Tamošauskas, G., Dubietis, A.: A compact, self-compression-based sub-3 optical cycle source in the 3−4 μm spectral range. J. Opt. **19**, 105505 (2017)
54. Lu, F., Xia, P., Matsumoto, Y., Kanai, T., Ishii, N., Itatani, J.: Generation of sub-two-cycle CEP-stable optical pulses at 3.5 μm from a KTA-based optical parametric amplifier with multiple-plate compression. Opt. Lett. **43**, 2720–2723 (2018)
55. Jargot, G., Daher, N., Lavenu, L., Delen, X., Forget, N., Hanna, M., Georges, P.: Self-compression in a multipass cell. Opt. Lett. **43**, 5643–5646 (2018)
56. Gholam-Mirzaei, S., Beetar, J.E., Chacón, A., Chini, M.: High-harmonic generation in ZnO driven by self-compressed mid-infrared pulses. J. Opt. Soc. Am. B **35**, A27–A31 (2018)
57. Garejev, N., Jukna, V., Tamošauskas, G., Veličkė, M., Šuminas, R., Couairon, A., Dubietis, A.: Odd harmonics-enhanced supercontinuum in bulk solid-state dielectric medium. Opt. Express **24**, 17060–17068 (2016)
58. Stepanov, E.A., Lanin, A.A., Voronin, A.A., Fedotov, A.B., Zheltikov, A.M.: Solid-state source of subcycle pulses in the midinfrared. Phys. Rev. Lett. **117**, 043901 (2016)
59. Grynko, R.I., Nagar, G.C., Shim, B.: Wavelength-scaled laser filamentation in solids and plasma-assisted subcycle light-bullet generation in the long-wavelength infrared. Phys. Rev. A **98**, 023844 (2018)

60. Alfano, R.R. (ed.): The Supercontinuum Laser Source. Springer (2006)
61. Ziane, O., Zaiba, S., Melikechi, N.: Continuum generation in water and carbon tetrachloride using a picosecond Nd-YAG laser pulse. Opt. Commun. **273**, 200–206 (2007)
62. De Boni, L., Toro, C., Hernández, F.E.: Pump polarization-state preservation of picosecond generated white-light supercontinuum. Opt. Express **16**, 957–964 (2008)
63. Gragson, D.E., Alavi, D.S., Richmond, G.L.: Tunable picosecond infrared laser system based on parametric amplification in KTP with a Ti:sapphire amplifier. Opt. Lett. **20**, 1991–1993 (1995)
64. Zhang, J., Zhang, Q.L., Zhang, D.X., Feng, B.H., Zhang, J.Y.: Generation and optical parametric amplification of picosecond supercontinuum. Appl. Opt. **49**, 6645–6650 (2010)
65. Bradler, M., Baum, P., Riedle, E.: Continuum generation in laser host materials towards tabletop OPCPA. In: Ultrafast Phenomena 2010, paper ME25 (2010)
66. Bradler, M., Riedle, E.: Continuum generation in laser host materials with pump pulse durations covering the entire femtosecond regime. In: Advanced Solid-State Photonics (ASSP) 2011, paper AMD4 (2011)
67. Calendron, A.-L., Çankaya, H., Cirmi, G., Kärtner, F.X.: White-light generation with sub-ps pulses. Opt. Express **23**, 13866–13879 (2015)
68. Çankaya, H., Calendron, A.-L., Zhou, C., Chia, S.-H., Mücke, O.D., Cirmi, G., Kärtner, F.X.: 40-µJ passively CEP-stable seed source for ytterbium-based high-energy optical waveform synthesizers. Opt. Express **24**, 25169–25180 (2016)
69. Schulz, M., Riedel, R., Willner, A., Mans, T., Schnitzler, C., Russbueldt, P., Dolkemeyer, J., Seise, E., Gottschall, T., Hädrich, S., Duesterer, S., Schlarb, H., Feldhaus, J., Limpert, J., Faatz, B., Tünnermann, A., Rossbach, J., Drescher, M., Tavella, F.: Yb:YAG Innoslab amplifier: efficient high repetition rate subpicosecond pumping system for optical parametric chirped pulse amplification. Opt. Lett. **36**, 2456–2458 (2011)
70. Riedel, R., Stephanides, A., Prandolini, M.J., Gronloh, B., Jungbluth, B., Mans, T., Tavella, F.: Power scaling of supercontinuum seeded megahertz-repetition rate optical parametric chirped pulse amplifiers. Opt. Lett. **39**, 1422–1424 (2014)
71. Gražulevičiūtė, I., Skeivytė, M., Keblytė, E., Galinis, J., Tamošauskas, G., Dubietis, A.: Supercontinuum generation in YAG and sapphire with picosecond laser pulses. Lith. J. Phys. **55**, 110–116 (2015)
72. Galinis, J., Tamošauskas, G., Gražulevičiūtė, I., Keblytė, E., Jukna, V., Dubietis, A.: Filamentation and supercontinuum generation in solid-state dielectric media with picosecond laser pulses. Phys. Rev. A **92**, 033857 (2015)
73. Jukna, V., Garejev, N., Tamošauskas, G., Dubietis, A.: Role of external focusing geometry in supercontinuum generation in bulk solid-state media. J. Opt. Soc. Am. B **36**, A54–A60 (2019)
74. Alismail, A., Wang, H., Altwaijry, N., Fattahi, H.: Carrier-envelope phase stable, 5.4 µJ, broadband, mid-infrared pulse generation from a 1-ps, Yb:YAG thin-disk laser. Appl. Opt. **56**, 4990–4994 (2017)
75. Indra, L., Batysta, F., Hříbek, P., Novák, J., Hubka, Z., Green, J.T., Antipenkov, R., Boge, R., Naylon, J.A., Bakule, P., Rus, B.: Picosecond pulse generated supercontinuum as a stable seed for OPCPA. Opt. Lett. **42**, 843–846 (2017)
76. Cheng, S., Chatterjee, G., Tellkamp, F., Ruehl, A., Miller, R.J.D.: Multi-octave supercontinuum generation in YAG pumped by mid-infrared, multi-picosecond pulses. Opt. Lett. **43**, 4329–4332 (2018)
77. Schmidt, B.E., Hage, A., Mans, T., Légaré, F., Wörner, H.J.: Highly stable, 54 mJ Yb-InnoSlab laser platform at 0.5 kW average power. Opt. Express **25**, 17549–17555 (2017)
78. Kasmi, L., Kreier, D., Bradler, M., Riedle, E., Baum, P.: Femtosecond single-electron pulses generated by two-photon photoemission close to the work function. New J. Phys. **17**, 033008 (2015)
79. Emons, M., Steinmann, A., Binhammer, T., Palmer, G., Schultze, M., Morgner, U.: Sub-10-fs pulses from a MHz-NOPA with pulse energies of 0.4 µJ. Opt. Express **18**, 1191–1196 (2010)

80. Fattahi, H., Wang, H., Alismail, A., Arisholm, G., Pervak, V., Azzeer, A.M., Krausz, F.: Near-PHz-bandwidth, phase-stable continua generated from a Yb:YAG thin-disk amplifier. Opt. Express **24**, 24337–24346 (2016)
81. Thiré, N., Maksimenka, R., Kiss, B., Ferchaud, C., Bizouard, P., Cormier, E., Osvay, K., Forget, N.: 4-W, 100-kHz, few-cycle mid-infrared source with sub-100-mrad carrier-envelope phase noise. Opt. Express **25**, 1505–1514 (2017)
82. Rigaud, P., van de Walle, A., Hanna, M., Forget, N., Guichard, F., Zaouter, Y., Guesmi, K., Druon, F., Georges, P.: Supercontinuum-seeded few-cycle midinfrared OPCPA system. Opt. Express **24**, 26494–26502 (2016)
83. Archipovaite, G.M., Petit, S., Delagnes, J.-C., Cormier, E.: 100 kHz Yb-fiber laser pumped 3 μm optical parametric amplifier for probing solid-state systems in the strong field regime. Opt. Lett. **42**, 891–894 (2017)
84. Neuhaus, M., Fuest, H., Seeger, M., Schötz, J., Trubetskov, M., Russbueldt, P., Hoffmann, H.D., Riedle, E., Major, Z., Pervak, V., Kling, M.F., Wnuk, P.: 10 W CEP-stable few-cycle source at 2 μm with 100 kHz repetition rate. Opt. Express **26**, 16074–16085 (2018)
85. Kanai, T., Malevich, P., Kangaparambil, S.S., Ishida, K., Mizui, M., Yamanouchi, K., Hoogland, H., Holzwarth, R., Pugzlys, A., Baltuska, A.: Parametric amplification of 100 fs mid-infrared pulses in ZnGeP$_2$ driven by a Ho:YAG chirped-pulse amplifier. Opt. Lett. **42**, 683–686 (2017)
86. Malevich, P., Kanai, T., Hoogland, H., Holzwarth, R., Baltuška, A., Pugžlys, A.: Broadband mid-infrared pulses from potassium titanyl arsenate/zinc germanium phosphate optical parametric amplifier pumped by Tm, Ho-fiber-seeded Ho:YAG chirped-pulse amplifier. Opt. Lett. **41**, 930–933 (2017)
87. Sandhu, A.S., Banerjee, S., Goswami, D.: Suppression of supercontinuum generation with circularly polarized light. Opt. Commun. **181**, 101–107 (2000)
88. Srivastava, A., Goswami, A.: Control of supercontinuum generation with polarization of incident laser pulses. Appl. Phys. B **77**, 325–328 (2003)
89. Vasa, P., Dota, K., Singh, M., Kushavah, D., Singh, B.P., Mathur, D.: Power- and polarization-dependent supercontinuum generation in α-BaB$_2$O$_4$ crystals by intense, near-infrared, femtosecond laser pulses. Phys. Rev. A **91**, 053837 (2015)
90. Kartazaev, V., Alfano, R.R.: Supercontinuum generated in calcite with chirped femtosecond pulses. Opt. Lett. **32**, 3293–3295 (2007)
91. Faccio, D., Averchi, A., Lotti, A., Kolesik, M., Moloney, J.V., Couairon, A., Di Trapani, P.: Generation and control of extreme blueshifted continuum peaks in optical Kerr media. Phys. Rev. A **78**, 033825 (2008)
92. Potemkin, F.V., Mareev, E.I., Smetanina, E.O.: Influence of wavefront curvature on supercontinuum energy during filamentation of femtosecond laser pulses in water. Phys. Rev. A **97**, 033801 (2018)
93. Ni, X., Wang, C., Liang, X., Al-Rubaiee, M., Alfano, R.R.: Fresnel diffraction supercontinuum generation. IEEE J. Sel. Top. Quantum Electron. **10**, 1229–1232 (2004)
94. Bradler, M., Baum, P., Riedle, E.: Femtosecond continuum generation in bulk laser host materials with sub-μJ pump pulses. Appl. Phys. B **97**, 561–574 (2009)
95. Jukna, V., Galinis, J., Tamošauskas, G., Majus, D., Dubietis, A.: Infrared extension of femtosecond supercontinuum generated by filamentation in solid-state media. Appl. Phys. B **116**, 477–483 (2014)
96. Schumacher, D.: Controlling continuum generation. Opt. Lett. **27**, 451–453 (2002)
97. Dharmadhikari, J.A., Dharmadhikari, A.K., Dota, K., Mathur, D.: Influencing supercontinuum generation by phase distorting an ultrashort laser pulse. Opt. Lett. **40**, 241–243 (2015)
98. Thompson, J.V., Zhokhov, P.A., Springer, M.M., Traverso, A.J., Yakovlev, V.V., Zheltikov, A.M., Sokolov, A.V., Scully, M.O.: Amplitude concentration in a phase-modulated spectrum due to femtosecond filamentation. Sci. Rep. **7**, 43367 (2017)
99. Romero, C., Borrego-Varillas, R., Camino, A., Mínguez-Vega, G., Mendoza-Yero, O., Hernández-Toro, J., Vázquez de Aldana, J.R.: Diffractive optics for spectral control of the supercontinuum generated in sapphire with femtosecond pulses. Opt. Express **19**, 4977–4984 (2011)

100. Borrego-Varillas, R., Romero, C., Mendoza-Yero, O., Mínguez-Vega, G., Gallardo, I., Vázquez de Aldana, J.R.: Femtosecond filamentation in sapphire with diffractive lenses. J. Opt. Soc. Am. B **30**, 2059–2065 (2013)

101. Kaya, N., Strohaber, J., Kolomenskii, A.A., Kaya, G., Schroeder, H., Schuessler, H.A.: White-light generation using spatially-structured beams of femtosecond radiation. Opt. Express **20**, 13337–13346 (2012)

102. Borrego-Varillas, R., Perez-Vizcaino, J., Mendoza-Yero, O., Minguez-Vega, G., de Aldana, J.R.V., Lancis, J.: Controlled multibeam supercontinuum generation with a spatial light modulator. IEEE Photon. Technol. Lett. **26**, 1661–1664 (2014)

103. Zhdanova, A.A., Shen, Y., Thompson, J.V., Scully, M.O., Yakovlev, V.V., Sokolov, A.V.: Controlled supercontinua via spatial beam shaping. J. Mod. Opt. **65**, 1332–1335 (2018)

104. Zhong, Y., Diao, H., Zeng, Z., Zheng, Y., Ge, X., Li, R., Xu, Z.: CEP-controlled super-continuum generation during filamentation with mid-infrared laser pulse. Opt. Express **22**, 29170–29178 (2014)

105. Wang, K., Qian, L., Luo, H., Yuan, P., Zhu, H.: Ultrabroad supercontinuum generation by femtosecond dual-wavelength pumping in sapphire. Opt. Express **14**, 6366–6371 (2006)

106. Kolomenskii, A.A., Strohaber, J., Kaya, N., Kaya, G., Sokolov, A.V., Schuessler, H.A.: White-light generation control with crossing beams of femtosecond laser pulses. Opt. Express **24**, 282–293 (2016)

107. Stelmaszczyk, K., Rohwetter, P., Petit, Y., Fechner, M., Kasparian, J., Wolf, J.-P., Wöste, L.: White-light symmetrization by the interaction of multifilamenting beams. Phys. Rev. A **79**, 053856 (2009)

108. Li, P.-P., Cai, M.-Q., Lü, J.-Q., Wang, D., Liu, G.-G., Tu, C., Li, Y., Wang, H.-T.: Unveiling of control on the polarization of supercontinuum spectra based on ultrafast birefringence induced by filamentation. J. Opt. Soc. Am. B **35**, 2916–2922 (2018)

109. Dubietis, A., Tamošauskas, G., Šuminas, R., Jukna, V., Couairon, A.: Ultrafast supercontinuum generation in bulk condensed media. Lith. J. Phys. **57**, 113–157 (2017)

110. Jimbo, T., Caplan, V.L., Li, Q.X., Wang, Q.Z., Ho, P.P., Alfano, R.R.: Enhancement of ultrafast supercontinuum generation in water by the addition of Zn2+ and K+ cations. Opt. Lett. **12**, 477–479 (1977)

111. Wang, C., Fu, Y., Zhou, Z., Cheng, Y., Xu, Z.: Femtosecond filamentation and supercontinuum generation in silver-nanoparticle-doped water. Appl. Phys. Lett. **90**, 181119 (2007)

112. Vasa, P., Singh, M., Bernard, R., Dharmadhikari, A.K., Dharmadhikari, J.A., Mathur, D.: Supercontinuum generation in water doped with gold nanoparticles. Appl. Phys. Lett. **103**, 111109 (2013)

113. Vasa, P., Dharmadhikari, J.A., Dharmadhikari, A.K., Sharma, R., Singh, M., Mathur, D.: Supercontinuum generation in water by intense, femtosecond laser pulses under anomalous chromatic dispersion. Phys. Rev. A **89**, 043834 (2014)

114. Wang, Y., Ni, H., Zhan, W., Yuan, J., Wang, R.: Supercontinuum and THz generation from Ni implanted $LiNbO_3$ under 800 nm laser excitation. Opt. Commun. **291**, 334–336 (2013)

115. Robinson, T.S., Patankar, S., Floyd, E., Stuart, N.H., Hopps, N., Smith, R.A.: Spectral characterization of a supercontinuum source based on nonlinear broadening in an aqueous K_2ZnCl_4 salt solution. Appl. Opt. **56**, 9837–9845 (2017)

116. Wang, L., Fan, Y.-X., Yan, Z.-D., Wang, H.-T., Wang, Z.-L.: Flat-plateau supercontinuum generation in liquid absorptive medium by femtosecond filamentation. Opt. Lett. **35**, 2925–2927 (2010)

117. Santhosh, C., Dharmadhikari, A.K., Alti, K., Dharmadhikari, J.A., Mathur, D.: Suppression of ultrafast supercontinuum generation in a salivary protein. J. Biomed. Opt. **12**, 020510 (2007)

118. Santhosh, C., Dharmadhikari, A.K., Dharmadhikari, J.A., Alti, K., Mathur, D.: Supercontinuum generation in macromolecular media. Appl. Phys. B **99**, 427–432 (2010)

119. Li, H., Shi, Z., Wang, X., Sui, L., Li, S., Jin, M.: Influence of dopants on supercontinuum generation during the femtosecond laser filamentation in water. Chem. Phys. Lett. **681**, 86–89 (2017)

120. Driben, R., Husakou, A., Herrmann, J.: Supercontinuum generation in aqueous colloids containing silver nanoparticles. Opt. Lett. **34**, 2132–2134 (2009)

121. Kulchin, Y.N., Golik, S.S., Proschenko, D.Y., Chekhlenok, A.A., Postnova, I.V., Mayor, A.Y., Shchipunov, Y.A.: Supercontinuum generation and filamentation of ultrashort laser pulses in hybrid silicate nanocomposite materials on the basis of polysaccharides and hyperbranched polyglycidols. Quantum Electron. **43**, 370–373 (2013)

122. Ramachandran, H., Dharmadhikari, J.A., Dharmadhikari, A.K.: Femtosecond supercontinuum generation in scattering media. J. Opt. Soc. Am. B **36**, A38–A42 (2019)

123. Paipulas, D., Balskienė, A., Sirutkaitis, V.: Experimental study of filamentation and supercontinuum generation in laser-modified fused silica. Lith. J. Phys. **52**, 327–333 (2012)

124. Šuminas, R., Tamošauskas, G., Jukna, V., Couairon, A., Dubietis, A.: Second-order cascading-assisted filamentation and controllable supercontinuum generation in birefringent crystals. Opt. Express **25**, 6746–6756 (2017)

125. Dubietis, A., Polesana, P., Valiulis, G., Stabinis, A., Di Trapani, P., Piskarskas, A.: Axial emission and spectral broadening in self-focusing of femtosecond Bessel beams. Opt. Express **15**, 4168–4175 (2007)

126. Sun, X., Gao, H., Zeng, B., Xu, S., Liu, W., Cheng, Y., Xu, Z., Mu, G.: Multiple filamentation generated by focusing femtosecond laser with axicon. Opt. Lett. **37**, 857–859 (2012)

127. Majus, D., Dubietis, A.: Statistical properties of ultrafast supercontinuum generated by femtosecond Gaussian and Bessel beams: a comparative study. J. Opt. Soc. Am. B **30**, 994–999 (2013)

128. Dota, K., Pathak, A., Dharmadhikari, J.A., Mathur, D., Dharmadhikari, A.K.: Femtosecond laser filamentation in condensed media with Bessel beams. Phys. Rev. A **86**, 023808 (2012)

129. Dota, K., Dharmadhikari, J.A., Mathur, D., Dharmadhikari, A.K.: Supercontinuum generation in barium fluoride using Bessel beams. Chin. J. Phys. **52**, 431–439 (2014)

130. Polynkin, P., Kolesik, M., Moloney, J.: Filamentation of femtosecond laser Airy beams in water. Phys. Rev. Lett. **103**, 123902 (2009)

131. Ament, C., Kolesik, M., Moloney, J.V., Polynkin, P.: Self-focusing dynamics of ultraintense accelerating Airy waveforms in water. Phys. Rev. A **86**, 043842 (2012)

132. Gong, C., Li, Z., Hua, L.Q., Quan, W., Liu, X.J.: Angle-resolved conical emission spectra from filamentation in a solid with an Airy pattern and a Gaussian laser beam. Opt. Lett. **41**, 4305–4308 (2016)

133. Neshev, D.N., Dreischuh, A., Maleshkov, G., Samoc, M., Kivshar, Y.S.: Supercontinuum generation with optical vortices. Opt. Express **18**, 18368–18373 (2010)

134. Maleshkov, G., Neshev, D.N., Petrova, E., Dreischuh, A.: Filamentation and supercontinuum generation by singular beams in self-focusing nonlinear media. J. Opt. **13**, 064015 (2011)

135. Aleksandrovsky, A.S., Vyunishev, A.M., Zaitsev, A.I., Slabko, V.V.: Random quasi-phase-matched nonlinear optical conversion of supercontinuum to the ultraviolet. Appl. Phys. Lett. **103**, 251104 (2013)

136. Suchowski, H., Porat, G., Arie, A.: Adiabatic processes in frequency conversion. Laser Photon. Rev. **8**, 333–367 (2014)

137. Krogen, P., Suchowski, H., Liang, H., Flemens, N., Hong, K.-H., Kärtner1, F.X., Moses, J.: Generation and multi-octave shaping of mid-infrared intense single-cycle pulses. Nat. Photon. **11**, 222–226 (2017)